R.E.GENERATION

R.E.GENERATION

Michael J. Prewer

MidnightSun

First published 2024 by MidnightSun Publishing Pty Ltd
PO Box 3647, Rundle Mall, SA 5000, Australia.
www.midnightsunpublishing.com

Cover design by Abby Stout
Internal design by Zena Shapter

Typeset in Copperplate and Adobe Casion.

Printed and bound in Australia by McPherson's. The papers used by MidnightSun in the manufacture of this book are natural, recyclable products made from wood grown in well managed forests.

NATIONAL LIBRARY OF AUSTRALIA

A catalogue record for this book is available from the National Library of Australia

FSC www.fsc.org MIX Paper | Supporting responsible forestry FSC® C001695

For

Elsie, who made me start writing.

And

Georgie, who supported me to the end.

SARAH

CHAPTER 1

The instant that I wake up, I know something is wrong. That something terrible has happened. Not because of a supernatural feeling, but because that's how my chemo-pod was programmed. At some point Earth died, but now the atmosphere is safe to breathe again. The toxicity has disappeared and Earth has been reborn.

I check the atmospheric data displayed on a monitor to my left. The oxygen levels are appropriate and no harmful gases are detected. The air temperature is twenty two degrees Celsius. I look to the digital clock which reads: 08:49, 27/07/53,219. Holy crap, it worked. I've been in here for more than fifty-thousand years.

A calm, recorded voice fills the pod. 'Welcome Sarah. The R.E.Generation Project wishes you all the best for your new life.'

Finding everything in order, I grab the handle next to my right arm and pull it down firmly. There is a whistling of gas as the airlock is broken. I place both hands against the lid of the pod and push hard. At first, the lid moves excruciatingly slowly and I grit my teeth with the effort but as I keep pushing, the momentum of the lid starts to take over and eventually it swings fully open. A huge cloud of dust billows up into the air and I cover my face from the falling debris. Once I can feel that nothing else is landing on me, I slowly sit up, take off my oxygen mask and look around. Nothing could've prepared me for what I see.

What was once a clinical room filled with eight chemo-pods is now a pine forest. Only minutes ago, I was sitting in my pod looking at the nervous and excited faces of my fellow volunteers.

Ang had given me an encouraging thumbs-up. He said, 'See you on the other side, Sarah,' before lying down and closing his lid.

Now, where his pod should be, all I see is a large tree trunk. His pod is simply gone. Four other pods are scattered around, however, only Karim's is where it should be. What I presume to be Meaghan's and Sam's pods are lying on their sides, and Kyle's is misshapen and squeezed between two trees. Like Ang's chemo-pod, Pam and Mabel's pods have simply vanished.

A wave of loneliness washes over me as I realise that no one else has reactivated, and that I may never see any of the others again. I feel numb, as my mind tries unsuccessfully to comprehend the huge jump in time that I've just experienced.

The scent of pine is strong in the air, but I breathe in a mouthful of dust which clings to the back of my nose and throat. I stand up carefully and walk over to Kyle's pod. The seal is broken and I glimpse white bone inside. I shudder and turn away in disgust. As a nurse, my job is to protect my fellow volunteers, but I wasn't able to save Kyle. I shake my head sadly. How could the life of such a vibrant young man be snuffed out so quickly?

I take a deep breath to calm myself, walk across to Karim's pod and wipe a thick layer of dirt and decayed plant matter off the metal shell. Everything seems to be as it should be, yet there's no sign of re-activation.

There should be three more rooms nearby, each with eight more volunteers in chemo-pods, however, looking at the carnage surrounding me, I realise I can't expect salvation from anyone else. Still, I have to have hope.

As I make my way towards where the other 'rooms' should be, I scan the forest around me. It's eerily silent. There are no bird calls. No insects buzzing around my head. Just a faint murmur of wind in the trees. The plants are familiar but not the native species that I would've expected. I am surrounded by large pine trees whose arms spread out overhead, producing a subdued shade. They tower above me, except for some shorter specimens that are suffocated by creepers. Some small, reddish-brown cones litter the ground. The undergrowth is sparse, allowing me to walk through the trees with relative freedom. The silence is overwhelming and the loneliness washes over me once more. Right now, I could feasibly be the only living creature on the planet.

I step over some crumbled mortar. A sad homage to the clinical laboratory that once housed this unit of the R.E.Generation Project. The next room contained the Alpha volunteers and I am surprised to find a small clearing with a half-ring of pods exactly where they should be. While they appear to be in pristine condition, the four pods remain firmly shut. I see three other pods underneath a fallen and decomposing log. Of the eighth and final pod, there's no sign. I pass through the clearing and keep moving, looking for the Gamma and Epsilon volunteers. I have plenty of day to work with, but I don't fancy the idea of a night in this forest on my own, even if there don't appear to be any animals around. Considering it's still early in the day, I am surprised by how warm it is as I weave my way between the trees. The air is thick with humidity and my skin glistens with sweat. When the laboratory was still standing, this was a cool, temperate environment all year round. The location was deliberately chosen for its high elevation and distance from the edges of the tectonic plates. Supposedly, the building was most likely to survive in this sort of remote, desolate, alpine environment. However, with significant climate change creating a new ecosystem and a large forest appearing from nowhere, the laboratory didn't stand a chance.

I almost miss the Gamma volunteers' pods completely. Only at the last minute do I notice a pod covered in pine needles jutting out of a mound of dirt. Looking around carefully, I struggle to make sense of what I'm seeing. I'm halfway up a small hill but the laboratory was originally built on flat ground. Trying to orient myself, I decide that some sort of landslide must've covered the remaining pods before the forest grew over it. I sigh in frustration and continue up the hill in the general direction of the Epsilon volunteers, though I expect them to be equally covered by the landslide. I reach the crest of the hill and as I survey the forest below me, a glint of metal catches my eye. I scramble down the slope and find an exposed Epsilon pod. It's covered in a pile of rubble from the collapsed roof, but otherwise looks largely intact. Looking around, I see two more pods half covered by the hill. The remaining five I have to assume are buried completely. Turning back, I shake my head, drumming my fingers on the exposed pod. Struggling to keep calm, I try to assess my options, but alone in a foreign forest, there's only one real option available to me. Survive. Survive and hope that someone else might wake up soon.

I turn and start to trek back up the hill. I have a medical kit, and some food in my pod, or at least I should. I'm not sure how long preserved food stays preserved. At the very least, I should have a decent supply of fresh water. I haven't seen any water sources on my outward trip, and I file away a memo to find a fresh supply as soon as possible.

Halfway up the hill, I stop in my tracks as I hear a whistling noise behind me. I smile at the unmistakable sound of gas escaping from an air lock. I turn around and race down the slope, excitement mixing with trepidation in the pit of my stomach. I reach the pod and hurriedly clear off all the debris. Plaster, mortar and rotten wood hit the forest floor and I see the lid begin to open slowly. Remembering the difficulty that I had opening my own pod, I help to lift it up.

Lying in front of me is a woman, I recognise her from training but being from a different group, I didn't have much to do with her. She looks at me, puzzled, probably trying to place my face too.

'When are we?' she asks.

'Fifty-thousand something,' I respond. 'Plenty has changed out here.'

She sits up. Looking around the forest, she nods her head, thinking. 'So, it worked then?'

'Yup. And I'm bloody glad to see you. Until now, I was the only one awake,' I say.

'Sorry, who are you?' she asks.

'Sarah. From Eta group. And you're Christine?'

'Cassidy,' she corrects. 'No one else has re-activated?'

'No. Not yet. But now that you're awake, maybe that means others will be waking up too.'

'It's so quiet,' she says serenely.

She stands up out of her pod, continuing to survey the forest. She wears the same practical outfit as me. Blue pants, a blue thigh-length jacket made from a synthetic material, and a green cotton undershirt. Her blonde hair, tied in a ponytail, falls halfway down her back and she stands tall and proud.

'It's just so weird, isn't it?' Cassidy ponders. 'This is exactly what we planned for, but it feels so sudden. I can't wrap my head around the time that we lost. It was then, and now it's now, and I did nothing in all the time in between. It's just so odd.'

'It is,' I agree. 'We trained for this, we thought about all the outcomes, but now we actually have to do it. I kinda thought that we would have to wait a long time. But *snap*, here we are. And we have to survive. No second chances. It starts now.'

'Do all of the groups look like this?' she asks. 'There's five pods missing and those two are half-buried.'

'Alpha looks the best. There were four pods that looked fine. Gamma was hardest hit. I only found one.'

'We should go to the Alpha group then. See if any of them have awoken. Should I grab my things?' Cassidy asks.

'What do you have?' I ask.

'A medical kit, fabric and sewing gear. Doctor and Seamstress. You?'

In addition to general survival training, all the R.E.Generation volunteers have training in at least two specialist fields. Medical, engineering, building, horticulture, zoology and hunting are identified as key skills for survival and therefore each group of eight volunteers has these skills amongst them. The secondary specialist areas are varied but identified as important for either survival or creating a new society.

'Nurse and Midwife,' I respond.

Cassidy nods acknowledgement. 'Can you lead the way to Alpha then?'

We walk in silence, the ground soft beneath our feet. The soil is loamy and covered in dry orange pine needles. Cassidy studies our surroundings and occasionally nods to herself. I'm not quite sure what to make of her. She seems serious and calculating, but I hope that we can be friends, especially if it ends up just being the two of us.

We pass the semi-submerged Gamma pod and Cassidy nods again. Suddenly, a loud call breaks the silence, coming from up ahead.

'Hello! Is anyone else here? Hello.' A male voice.

'Hello!' I call back. 'We're coming, stay where you are.'

'Hello!' Another voice calls, further off. Karim's perhaps.

Cassidy and I glance at each other and break into a jog. We emerge into the clearing and I see a young man sitting on his pod. He has slick dark hair and his posture exudes confidence.

'Names and roles?' he asks.

'And hello to you too,' I say. 'I've just been walking through a forest that used to be clear tundra, but you're right, straight to business. Sarah, Nurse and Midwife, I also–'

'Okay,' he says, cutting me off. 'You?'

'Cassidy. Doctor and Seamstress.'

'Better than another nurse, I guess. But we're gonna need more than just medical if we're gonna make anything of this.'

'And you are?' I ask, trying to ignore his arrogance and complete lack of respect for me.

'Declan. Doctor and Dentist.'

I manage to stop myself from rolling my eyes and smile politely instead. Doctor and dentist, a suitably cocky combination.

We hear another Alpha airlock open and at the same time, Karim emerges from the trees on the other side of the clearing. He has tightly coiled black hair and a broad nose. I run towards him and wrap my arms around his middle, thrilled to see a familiar, friendly face. Karim is a good foot taller than me, so my head barely reaches his chest, but he holds me in a warm embrace.

'Hi, Sarah.'

'Name and–' Declan starts to ask, but I cut him off almost immediately.

'You're incorrigible, Declan. Is that really the way you're going to greet anyone who wakes up? But this is Karim. Engineer and Accountant, if you must know.'

'What type of engineer? If I may be so bold,' Cassidy asks.

'Of course you may,' Karim says, smiling winningly at Cassidy, his teeth a brilliant white in contrast to his dark skin. 'Electrical. And you are?'

We repeat introductions again and again as other Alpha members re-activate. Natalie, carpenter and chef, makes a point of shaking everyone's hand and squeezes mine with unnecessary force. Ayesha, botanist and teacher, has attractive features, tanned skin and long flowing hair, but is reasonably shy and hangs back from group discussions. Tomohiro is a charmer, enquiring after the health of all the women before allowing any other discussion. He's a trained blacksmith and geologist, but he assures us that he also has many other skills and experiences, including time as an arborist, going down mines and tearing up the dance floor.

As a group, we revisit each of the sites, but none of the other chemo-pods seem to be stirring to life.

No one has spotted any animal life, so finding food is an imperative, even though our dry supplies should last at least a month. We also need to find water and build a shelter. The temperature is rising rapidly and although the trees offer some shade, they won't protect us if it rains.

Natalie and Tomohiro take charge building a shelter. Tomohiro has an array of saws that he, Karim and Declan use to fell trees and Natalie has carpentry tools to whittle and shape the wood. Ayesha stays with Natalie, helping where she can, making notes in a small notebook about the plants she encounters.

Cassidy and I grab some rucksacks and a compass and strike out in a north-easterly direction, looking for food and water. Over the last hour, Cassidy has relaxed, and we chat while we walk.

'Declan is a piece of work,' I comment. 'He didn't seem too bad the few times I bumped into him at training, but the instant we got here he turned into an arrogant prick.'

Cassidy nods. 'Sharon, one of the other Epsilons, told me to be careful around him. Apparently, he was originally from money, but his dad disowned him after an incident in his final year of school. He threw himself into his medical studies, but they never made up.'

'What was the incident?' I ask.

'No idea.'

We find some fungi in a particularly shady section of forest.

Large, plump mushrooms, bigger than a dinner plate. The top side is orangey brown, and the underside is yellow and spongy. We harvest one mushroom, slice it in half and put it in a bag for Ayesha to analyse, hoping it's edible. After bagging the fungus, I notice that there are many bugs and grubs burying into the overturned earth. It's a relief to see something else alive, even if they're just insects.

I point them out to Cassidy who beams in excitement.

'So not all animal life was killed after all,' she says. 'And I'd say, where there's insects, there's probably something that eats them.'

'Not necessarily,' I reply hesitantly.

We dig in the earth and select some of the largest grubs, putting them in a separate bag. Resembling witchetty grubs, there appear to be two distinct species. A smaller, cream-coloured grub with brown and orange stripes and a large yellow grub with one bright blue line running the length of its body. Some of this type are bigger than my thumb and will provide good nutrition if they're not poisonous.

'I wish Ang was here. He'd probably be able to tell us whether they're safe to eat,' I say.

'Or Newton. He could've done a chemical analysis,' Cassidy adds.

'Ang's pod was just gone. How could it be gone?' I ask.

'Thousands of years have passed, Sarah. Anything could've happened.'

'But the pods were built to last. Look at mine and yours.'

'But the tech isn't perfect. So far, seven of thirty two pods made it. There were always going to be things they couldn't account for.'

'We didn't expect a bloody forest to grow, did we? It was weird opening my pod and seeing a tree where Ang had been only moments before.'

'Did you look in the tree?' she asks.

'What?'

She gestures to the canopy above us. 'Well, if Ang's pod was replaced by a tree, maybe he's up there? People never think to look up.'

'Maybe.' I adjust the bag of fungus that's working its way off my shoulder. 'I'm not looking forward to investigating the other pods closely. We don't know what we'll find.'

'Well, Newton's pod was next to mine.' Cassidy says. 'It was partially buried, but I'm hoping he's still in there.'

'Kyle's was definitely cracked open,' I add glumly.

'The one between the trees in your group?' she asks.

'Yeah. Now it's just his skeleton inside. He was nice.'

'So were the other Epsilons. Who knows how the buried pods are?' she says, shaking her head.

We move back towards the group in sombre silence. Both saying farewells in our heads, without knowing exactly who to farewell.

Inattentive in my subdued state, I almost trip over a large brown object lying on the ground. We stop and, bending to inspect it, realise it's an old mushroom. It's dried out and doesn't look very appetizing, but it's clear that flesh was torn off one side when it was still fresh. We've travelled more than a hundred metres from where we found the other mushrooms and we look at each other, startled. Cassidy is the first to voice her worries.

'An animal big enough to carry this here is big enough to attack humans.'

'And those are large chunks that have been torn off,' I add.

Cassidy inspects the damage closely and points to some grooves in the dried husk. 'Teeth marks.'

I pick up the dried mushroom and we run back towards the

group, furiously scanning the forest around us for any sign of danger. The dark brown trunks and drooping pine branches suddenly feel like they are closing in on me. I can hear my blood pumping in my ears and fear ripples through my body as I realise just how much we still have to learn about the world that we've been parachuted into.

ANG

Ang pushes through the crowd, jostled by the placard-holding 'Pro-Lifers' who heckle the volunteers as they arrive for training.

A man grabs Ang by the arm, falling to his knees beside him. He looks up at Ang with anguish in his eyes.

'Don't do this, young man. It's suicide. The Lord will provide!'

Ang pulls his arm clear and hurries, head down, for the front door of the complex. The Pro-Lifers have been camped outside the R.E.Generation headquarters for the last month and Ang is used to them by now, but they still have a way of getting under his skin. Ang sees one of the Beta volunteers wrapped in an embrace with a sobbing woman. As he pulls away from her, she calls out.

'What witchcraft have you used to turn my own son against me. Give him back to me!'

Ang is glad that his parents are supportive of his decision. As much as it pains them, and him, that he is heading towards a point of no return, they are fully aware of the state of the world. As climate scientists themselves, they've been monitoring the world for years as it hurtles towards oblivion. They've watched as humans squabble amongst themselves rather than implementing any meaningful changes.

Inside the complex, Ang shuffles towards the Eta group training bay. Today was not supposed to be the final day of training. They were supposed to have more time than this. His insides writhe with complex, competing emotions. Excitement at the thought of a new world, trepidation in trusting the R.E.Generation technology and system, and sadness at leaving his family behind.

He spots Sarah and gives her a quick wave. She smiles back and another emotion tries to force its way to the surface. There's still too much to get through first, but if they both make it, then she is the one that he wants to get to know better.

The day consists of final briefings, running over key information one last time. How to operate the chemo-pods, how to find safe drinking water and how reproduction amongst the volunteers needs to be organised to avoid genetic conflagration. They also memorise the locations of every preservation laboratory within five-hundred kilometres of their own, so that survivors can search for each other and create an ever-bigger society once they've got themselves established.

That night, Ang lies in bed unable to sleep, and unable to think of anything productive to do. Everything has been organised for him. His pod has been stocked with food, water, spare clothes and trade specific tools. His outfit for the coming day is laid out beside his bed. A plain blue pair of pants, a pale green shirt, socks, underwear and a blue slim-line jacket that falls halfway down his thighs. The pants and jacket are made from a specific hybrid fabric that is light, breathable and waterproof. After a few fitting sessions, they have been tailored to fit him perfectly. He also has a pair of lightweight boots and a sun hat. Despite the fact that their preservation date has been brought forward significantly, everything he needs is ready to go.

He gets out of bed and picks up a picture from his graduation.

His parents stand either side of him, beaming with pride. He takes the picture out of its frame and puts it inside the pocket of his neatly folded shirt. His brain still whirrs as he lies back down and it's hours later when he finally falls asleep.

After a quick breakfast, Ang's mother drives him back to the complex. The group of Pro-Lifers looks bigger than ever. Instead of heading inside, he stays in the car. He shares many sideways looks with his mother, but not a single word needs to be said.

A few minutes later, a large bus arrives. Ang and his mother get out of the car and as the Pro-Lifers start to surge towards them, they share a quick embrace. They break apart and his mother looks at him with sad eyes.

The Pro-Lifers hurl threats and warnings as they rush forward, but Ang doesn't hear a word. Although she is quieter than the rest, the only voice he can hear is his mother's as she quietly sends him on his way.

'We're so proud of you,' she says calmly.

Ang makes it onto the bus before any of the protesters get to him, but many of the other volunteers are not so lucky. They have to push through the crowd and are only able to relax once they've passed the security guard who is preventing the Pro-Lifers from clambering onboard.

The volunteers who trained at this complex in eight groups were divided into two further teams the previous day. Ang watches as the other Eta, Alpha, Gamma and Epsilon volunteers climb onto his bus. The Beta, Delta, Zeta and Theta volunteers board a neighbouring coach which is destined for a different destination.

On board, the Alpha volunteers monopolize the back seats and all sit together. They seem to defer to Declan, a well-built, dark-haired man who reserves seats and ushers the others to the back. Ang and Declan have chatted a few times previously during training and Declan has always been friendly and courteous. He seems well suited to the R.E.Generation Project. Driven and calm.

The Gamma, Epsilon and Eta volunteers sit in pairs but otherwise are intermingled. Pam sits beside Ang and smiles, but Ang sees the fear in her gaze. They chat quietly as the bus pulls away, leaving the Pro-Lifers and the R.E.Generation training facility behind.

'What are we doing?' Pam asks nervously. 'We didn't even do all the training they had planned.'

'It'll be fine. It'll be fun!' Ang responds, trying to sound confident.

'Fun? Fun! You call starving and struggling to survive fun?'

Pam continues to freak out, and Ang unsuccessfully tries to reassure her. Luckily, Sarah leans across the aisle and places a hand on Pam's leg.

'Pam, listen to me,' she says. 'Yes, it's scary. Yes, we don't know exactly what to expect and that's making us all nervous. But we're prepared as best we can. The Project has modelled countless futures and planned for as many as possible. Besides, we're in this together. We'll look out for you, and we know you'll look out for us.'

Pam gives Sarah a wan smile and takes a few deep breaths. Some of the nearby volunteers who had stopped to listen to Pam and Sarah return to their own conversations.

Ang and Pam sit in silence for the rest of the journey. Ang watches the scenery through the window as the bus winds its way upwards, higher into the mountains. At a certain point, they emerge above the tree line and he can see across the range. See the other mountain peaks and the vast valleys between. Far below, he can see a large section of forest that is slowly recovering after being decimated by a recent bushfire. An eagle circles in the sky overhead and a few kilometres away, further up the mountain, he spots a white building surrounded by a high fence and tundra. He notices a few cars parked outside the perimeter, and people holding placards. Some of the Pro-Lifers have made the trip out to this remote preservation lab.

A few minutes later, they pass the Pro-Lifers on the way into the fenced compound, but Ang pays them no heed. Unlike the training complex, this lab is designed for one purpose only. To house the dormant volunteers for the duration of their chemopreservation. The location was chosen for its geo-stability and predictable weather.

Once inside the building, they are met by a team of scientists who hurry them towards their pods. They are separated into their four groups and each directed towards a different wing of the building. Each pair of volunteers has a scientist who stays close and quietly directs the volunteers to their individual pods. There is no time for goodbyes or conversation. Ang wonders whether the scientists are deliberately hurrying and controlling the situation to prevent any of the volunteers from changing their minds.

The scientist allocated to Ang, a sensible looking woman with spectacles and a high ponytail, gives him his final instructions and tells him to lie down. Ang glances around the room one final time and makes eye contact with Sarah. He smiles at her and gives a quick thumbs up.

'See you on the other side,' he says.

Ang lies down in the pod and the scientist places an oxygen mask over his mouth and nose.

Ang pulls the lid closed and glances at the monitors lining the internal walls. A large panel above him recedes revealing a metal apparatus and thousands of tiny needles. He takes a deep breath, inhaling a strong anaesthetic, and a few moments later everything goes black.

As the scientist watches Ang lower the lid of the chemo-pod, she thinks about the accelerated nature of the process and wonders whether this youthful boy will get the future that he wants. They've made more compromises in the last few weeks than he or the other volunteers could possibly know.

SARAH

CHAPTER 2

Cassidy and I veer slightly off course in our mad rush back to camp, but the sounds of sawing and hammering lead us back to the others.

We find them in the Alpha clearing, where at least five more trees have been felled. A further two trees have had their tops lopped off about two metres from the ground. Natalie has turned one of the fallen trees into a large beam and has secured it on top of the two tall stumps. She splits sections of felled tree to make half-log planks, which she and Ayesha lean up against the beam to make a large, tent-like structure. Tomohiro and Declan are working on cutting down another tree.

Karim is resting on one of the smaller stumps. Seeing us covered in sweat and breathing hard, he raises a questioning eyebrow at us. 'Everything alright?'

'Get the others,' I pant back.

Karim tells the others to stop what they're doing and calls them over.

'This better be bloody important,' I hear Declan mumble.

I hold out the dried mushroom so they can all see it. 'We were wrong about there being no animals.'

'We know. We found a snake in one of the trees we toppled,' Tomohiro says.

'Pretty thing. Greeny-black with a yellow belly,' Ayesha adds.

'We killed it,' Declan says, blunt and to the point, as always.

'Well, this mushroom has been eaten by something bigger than a snake. Look at the teeth marks,' I say.

'And it carried this whole thing a fair way,' adds Cassidy.

'What made it leave it?' Karim asks.

'Maybe it was full?' Natalie suggests.

'Maybe it was scared off by something even bigger,' Tomohiro adds.

'Let's hope not,' I reply. I look around the group, seeing a mixture of interest and fear. 'I reckon our pods are the safest place for us to sleep tonight. At least until we know more about what's around.'

'They have kept us safe for years–' Ayesha begins.

'But that was when we were in stasis,' Natalie says. 'I don't think the pods were designed for staying in. Wouldn't we run out of oxygen?'

'Only if we seal the airlocks,' I say.

'If we don't seal the pods then wouldn't the creature be able to open them?' Tomohiro asks.

'The lids are heavy,' Ayesha says.

'Can we lock the pods without sealing them?' Cassidy asks.

Karim puts his hand up sheepishly. 'Pretty sure there is spare oxygen in the tanks–'

'Enough!' Declan looks around the group with contempt. 'You're all being daft. We're building the shelter, so we stay in the shelter. If we set a watch tonight, we'll be fine.'

I look around the group. Declan's scorn appears to have cowed most of the others. No one appears to want to speak up and contradict him. He is certainly imposing himself as a leader, although he's not the kind of leader I'd like to follow.

‘Why don’t we vote on it?’ I ask. ‘Those in favour of staying in the pods and those in favour of sleeping in the shelter.’

Natalie, Tomohiro, Declan and Cassidy vote for the shelter, so we agree to Declan’s plan. We eat rehydrated food for lunch until we can go through the process to decide whether the foods we have collected are safe for consumption. The first step is to rub each food on someone’s wrist and wait for a reaction.

We have four different foodstuffs to try: two grubs, the mushroom and the snake.

We decide that it’s best if Declan, Cassidy and myself stay away from the new foods, so that we can lend medical assistance if necessary. We fetch our medical supplies and bring them to the shelter, while the others decide who is going to try what.

When I show the grubs to the group, Ayesha raises concerns over the larger one. She thinks that the blue stripe indicates it’s likely to be poisonous. Declan points out that the orange and brown striped grub may also be toxic. Ideally, we would’ve had a chemist to do some rudimentary analysis on the food. But in the absence of that, we have to trust the process.

Karim and Tomohiro agree to play rock, paper, scissors to decide who will try which grub, however, after choosing the same option four times in a row, Tomohiro throws his hands in the air and declares that he'll take the yellow and blue one.

Ayesha scratches the flesh of the mushroom and tells us that if it changes colour, it could indicate that it’s poisonous. Nothing happens, and because we know that an animal has eaten it, we hope that means it’s also safe for human consumption.

Nat is left with the snake, so she cuts off the tip of its tail and rubs the exposed flesh on her wrist. Tomohiro and Karim cut small slices off their grubs and follow suit, while Ayesha rubs mushroom onto her skin.

We work through the afternoon finishing the shelter, while

monitoring Nat, Tomohiro, Karim and Ayesha for symptoms. We all abandoned our jackets early in the day and it isn't long before the boys strip off their t-shirts too as we work up a sweat in the hot and humid conditions. Half an hour later, I take off my own shirt, leaving my bra on for modesty, and the other girls quickly follow suit. The shelter is a rudimentary design, but Nat tells me it's just a temporary, quick solution. Once it's finished, we step back to admire our handiwork.

Wooden walls enclose the little building on three sides in an A-frame design. The central beam is about four metres long, and half-log planks slant down on either side to make the main walls. Cassidy has attached a temporary triangle of fabric, split down the middle, over the final open side to create an entrance. Inside, there is head room under the beam itself, but the walls slant rapidly, meaning there is limited room for standing. When lying down there is ample space for all of us and we can also sit comfortably in an oval.

We are famished after working all day, so it isn't long before the conversation turns to dinner. We will all have to eat tasteless sludge, which was apparently once potato and beans, but the others will also get their first taste of the new food.

As a qualified chef, Natalie has some knives and pots amongst her belongings and she takes pleasure in filleting the meat of the snake and frying a small piece on a fire that Karim makes for her. There's plenty of dry kindling around, and we all have flints in our survival kits.

Once the food is ready Nat, Tomohiro, Karim and Ayesha look at each other, holding their respective morsels in front of them.

'Here goes nothing,' Karim says, and puts the grub in his mouth.

The others follow suit, chewing on the food before spitting them out onto their plates.

'How was it, Nat?' Declan asks.

'Mm, tasty,' Natalie replies, licking some meaty juice off her fingers.

'Mine was a bit tasteless,' Ayesha adds, looking a bit miffed.

'See, that wasn't so bad, was it?' Tomohiro smiles. 'I guess we're all gonna be fine.'

We all laugh, knowing that we'll have to wait longer before we know whether the food is safe. But as we laugh, I notice that Tomohiro's smile starts to become lopsided. A bemused look crosses his face and he touches his left cheek, where the muscles appear to have stopped working.

'Tomohiro!' I scream.

He spits on the ground, trying to get rid of any remnants of the grub that might still be in his mouth.

'Tomohiro, are you alright?' Ayesha asks.

Cassidy has moved to his side and is examining him.

'My face,' he says, his voice slurred. 'And, and my arm.'

'What do you mean?' Karim asks.

'My left arm. I can't lift it.'

All the muscles on the left side of his face have gone lax, causing his eye to droop closed. When he talks, only the right side of his mouth moves.

'You're gonna be alright,' Cassidy says soothingly.

'You can't say that,' Declan interjects. 'For all you know, all of the muscles in his body are gonna seize up and he won't be able to breathe.'

Cassidy glares at him. 'Gee, thanks Dec.'

'The symptoms set in fast, but they seem to have stopped worsening,' I say. 'Hopefully, this is as bad as it gets, Tomohiro.'

'Yeah. I feel great!' Tomohiro gives a thumbs up with his right hand. At least he's able to treat such a serious event with some humour.

'Can you wiggle your toes for me?' Cassidy asks. I notice that she is now sitting behind Tomohiro, propping him up. By the way she is straining I can tell that he's not supporting his own weight anymore.

We see Tomohiro wiggle the toes of his right foot and he also flexes his right ankle, but his left foot doesn't move at all.

'Why don't you lie down, Tomohiro. Have a relax,' she says calmly.

Cassidy helps Tomohiro lie down and he gives us another thumbs up. 'If I'm like this from now on, then I demand we set up our society with equal rights for the disabled, as well as a healthy pension.'

I can't help but laugh.

We decide on a roster for the night where Declan, Cassidy and I take turns to be on watch. Guarding against any strange beasts that may appear from the wilderness. Watching over Tomohiro to make sure he doesn't deteriorate, and also monitoring Karim, Nat and Ayesha to ensure they don't develop symptoms of their own. We have cut a few wooden spears which are placed near the doorway, ready to hand.

Cassidy takes the night's first sentry shift and when she wakes me, reports that nothing has changed. I clamber onto my knees, shaking my head to wake up properly.

I check on Tomohiro, but he appears to be sleeping peacefully so I decide not to wake him.

Inside the hut, with the warmth created by the other campers and the rhythmic sounds of their breathing, it's not long before I start getting drowsy. I head outside, hoping the fresh night air will help me stay awake.

The air is less stale, but not much cooler. The trees appear to be holding in the day's heat.

I lean against one of the walls, marvelling at Natalie's simple but effective handiwork. The stars provide a little light, but the forest is dark and menacing. The embers of our small fire glow in the gloom.

I spend the next three hours listening out for any disturbances in the surrounding forest and from within the hut, but nothing interesting happens. I check on Tomohiro periodically. His heart rate and breathing are normal, and he doesn't appear to have a temperature.

At the end of my shift, I wake Declan up, but he waves me off.

'Fuck off, Sarah. Let me sleep.'

'It's your shift, Declan,' I whisper, trying not to wake the others.

'I don't want to. You do it,' he says grumpily.

'I need sleep. And someone needs to keep an eye on Tomohiro and the others.'

'You're the nurse, you fucking do it,' he says, after which he rolls over and refuses to be roused.

I head back outside and pace around the hut to try and dissipate my frustration with Declan. We all completed medical and psychological assessments before we were accepted as volunteers. I'd hoped that would mean that everyone in this new world would be a nice person, but Declan is a prick, plain and simple. Maybe he fudged his psych assessment.

I keep circling the hut, as the movement helps me to stay awake. Not long after, I hear stirring inside and, stifling a yawn, I hope that Declan has decided to take his shift after all.

I move to the fabric flap and am almost bowled over by Karim as he stumbles out past me. He manages to move a few more metres away before he is violently ill at the base of a tree. I cross to him, placing a hand on the small of his back.

'You alright?' I ask.

He is leaning against the tree, his head down. He shakes his head, the movement bringing on another bout of sick.

'Water,' he croaks.

I find a canteen and bring it to him. After rinsing out his mouth, he attempts to drink a little. A few minutes later he heaves it up again. Supporting his weight, I help him to one of the exposed tree stumps and he sits down, his head in his hands. I feel his forehead, which is hot, and his hands are warm and clammy.

'How do you feel?' I ask.

'How does it look?'

'Sorry, I mean, apart from the vomiting. Do you have any pain? Any other difficulties?'

He goes to shake his head again but stops himself as he fights back a wave of nausea. 'No, otherwise I'm mostly fine. I just feel hot and sick.'

I find some cloth off-cuts from Cassidy's work and, after wetting them, place one on his forehead and one on the back of his neck. He attempts to have some more water and this time appears to keep it down.

I check on Tomohiro again, but his condition hasn't changed. I also check on Ayesha and Natalie more closely, but they both appear totally fine. Karim retches again and I go back outside to sit with him.

Over the next few hours, he starts to feel less hot, and he is slowly able to keep some water down.

As dawn breaks Ayesha emerges from the hut. 'You look terrible, Karim.'

I fill her in on his condition and she agrees to keep an eye on him so that I can get some sleep.

'Why are you up, anyway?' she asks. 'Did you and Dec swap shifts?'

'No. He refused to get up, so I did a double shift,' I say. 'And so, to bed.'

I sleep restlessly at first and every time the others get up, they interrupt my shallow sleep. There's a commotion as someone helps

Tomohiro to get outside. Finally, after they've all risen and left the hut, I manage to catch up on some sleep.

I wake again to the sound of sawing and hammering but continue to nap some more. By the time I actually get up, the sun is already high in the sky. I look around the little clearing and see the others at work.

Natalie is almost finished constructing another hut. Tomohiro is propped up next to a tree offering advice and Karim is sitting in the shade, his head in his hands. That leaves Declan, Cassidy and Ayesha to help with the construction, and from the language spurting from Nat's mouth, I judge that none of them are very handy with a saw.

I sit down next to Karim, who looks up as he hears me approaching.

'You alright?' I ask, before realising I've asked another stupid question.

'Imagine the worst hangover you've ever had,' he replies. 'Now multiply that by five and you might feel like I've felt this morning.'

'Can I do anything to help?' I ask.

'Nah. I'm feeling a bit better now. I just have to remember not to move too quickly. Or preferably, not move at all.'

I move across to Tomohiro and squat next to him. 'How do you feel?'

'Like we should've listened to Ayesha. My whole left side's gone.' His words are still slurry. His left eye weeps a thin fluid and his arm hangs limp at his side.

'I'm sorry,' I say. 'We shouldn't have let you eat that grub.' I feel the guilt deep in the pit of my stomach. I found the grub, so in a way, that means I poisoned Tomohiro.

'It was my choice. Just sucks, huh?'

I nod in agreement.

'Here in this brave new world and I'm gonna be a cripple, a liability,' he says shrugging his right shoulder.

'Don't you dare start thinking like that. You've still got knowledge that we'll count on to survive. You'll learn to do things with one hand. You can still write.'

He smiles at me. 'Thanks, Sarah.'

I join the others. They're glistening with sweat, the hot, humid weather and the work pushing their limits.

'Looking good,' I say, admiring the second building. Nat is following the same design as the initial hut. Two trees have had their tops chopped off about two metres from the ground and the central beam runs between them, resting on the elevated stumps. They have also put the half-log planks into place to make the diagonal walls. Declan is splitting some logs and Ayesha and Cassidy are helping Natalie build the third wall. They have a row of split logs laid out on the ground, forming a sort of triangle. Ayesha is hammering them together using a cross brace and Nat is measuring the gap it will fill. Once Ayesha is finished, Nat measures the wood, marks the correct dimensions and saws off the excess pieces. It's slow work, but she won't let any of us sub in.

Once it's finished, I help them lift the wall up under the A-frame and Natalie secures it into place with a few well-placed nails.

We repeat the process on the other end, but with two smaller triangles set apart to make an entryway.

'Phew,' Nat says when the final piece is in place. She wipes sweat from her forehead. 'I thought we'd never get this one finished.'

'Lunch?' Cassidy asks.

The weary workers sit with Karim and Tomohiro and we all eat rehydrated sludge.

'Man, I have a ripper headache,' Declan says.

'Ha ha, very funny,' Karim retorts. His forehead is no longer covered in sweat and he's looking more alive.

'No, I mean it.' Declan says.

'Me too,' Ayesha says. 'You know what it is though?'

Declan shake his head. 'Nuh. Do tell.'

'No coffee!' she says. 'I just feel like I need an espresso.'

'Oh man, why'd you have to say that?' Tomohiro says. 'Now all I want is my regular from Carrie's Cafe. They had the ultimate coffee.'

'No coffee. Yup, that's it. I've got a coffee headache,' Declan says.

'Better get used to it,' Nat says. 'You aren't gonna get any coffee around here.'

Ayesha, Tomohiro and Declan moan in unison.

'Who's gonna sleep in which hut?' Cassidy asks, changing the subject to take the others' minds off their coffee withdrawal.

'Boys and girls,' I suggest. That seems most logical.

'I think we should divide by volunteer group,' Declan says. 'Alphas in one hut and you guys in the other. There's only three of you anyway.'

I've got nothing against staying with Karim and Cassidy. They're both lovely people. And even more so I'd be thrilled to be away from Declan. But I don't like the idea of dividing by group. We're all equal here and there's no point creating sects or building arbitrary walls. Moreover, it goes against what we were taught in training about group governance.

'That doesn't make much sense,' I say. 'Our training groups don't mean anything here, besides–'

'Let's vote on it,' Declan says, cutting me off. 'Those in favour of housing by training group?' Declan raises his hand and Natalie, Ayesha and Tomohiro soon follow. 'That's four to three, matter solved.'

Declan's confidence in going straight to a vote and the speed at which the other Alphas followed his lead makes me suspicious. Almost like they'd already discussed it in private. I realise there are already power structures at work here and games being played behind the scenes. Maybe sleeping most of the morning wasn't the best move after all.

What I can't understand is what hold Declan has over the other Alphas. What makes them follow his lead so readily?

'Okay, I guess that's that,' I say diplomatically. Clearly there is a lot about our training that Declan is going to ignore.

'What's the plan for the rest of the day?' Karim asks.

'I'm gonna build the shit house,' Nat says.

For the last twenty four hours, we've all been doing our business out in the forest, but that's hardly sustainable.

'Declan, you can help dig the pit, yeah?'

Declan smiles at Nat. 'Of course.'

'There's a small shovel in my pod, if you need it,' Tomohiro adds.

'We need water,' Karim says. 'We can't survive on our supplies forever.'

'And we should get more of that fungus,' Ayesha adds.

'Cassidy, Ayesha, Sarah, you go scouting then,' Declan says, taking control. It also doesn't escape my notice that he puts me last. 'We'll keep an eye on the boys and keep building.'

As we trudge into the forest, my head starts to throb. I may have slept in later than the others, but I, too, am used to my morning coffee, and this is day one without it.

TESS

'Welcome to *Tomaten Van Morgan*, one of the world's largest tomato growers,' Tess says, beginning her introduction for the new tourists. 'And as you can see, every single tomato we grow is produced indoors in these massive greenhouses.'

They stand at the start of a row of tomatoes that they cannot see the end of. They stroll past countless rows of plump red fruit and the glass roof soars above their heads.

'Why do we grow our tomatoes in greenhouses?' Tess asks. 'Well, I'm sure that, today, the answer is obvious.' She gestures to the roof where torrential rain lashes against the glass.

'No matter what the weather is outside,' she calls, fighting to be heard above the noise created by the rain. 'We can grow perfect tomatoes in here. The temperature and humidity are kept at a constant level and even on days like today, when there is no sun, we can provide the UV to help the plants grow and the tomatoes ripen.'

The tourists nod and look suitably impressed.

Tess takes a sip of coffee while the tour members have a look around. She forgot her Keep Cup today and feels guilty about the disposable one in her hand, but nothing can get between Tess and her morning brew.

'Would anyone like to try a tomato?' Tess asks.

There are many nods, but Tess isn't even looking, assuming that everyone will want to try some. She takes out a knife and hacks off a cluster of tomatoes.

'I can promise you that this will be the nicest tomato you have ever tasted,' she says, beaming.

She hands out the tomatoes and everyone enjoys the fresh produce. There is giggling as one tourist manages to get tomato juice and seeds all over their face.

Tess, are you there? Zoë asks, the message being relayed directly into Tess's EarBud.

'Yes, I am. Go ahead,' she replies.

I need you to come to Greenhouse 13, Sector B.

'I'm in the middle of a tour. Can Mila come?'

Fuck. Get the tour back to the main building. It's not safe in the greenhouses.

'What do you mean?'

Just get them into a building and then come meet me. Mila is already on her way.

'I'm so sorry to bring the tour to an abrupt halt ladies and gentlemen, but I have just been informed that it is currently unsafe in the greenhouses and I need to get you to safety. If you will all please follow me.'

She leads them quickly back to reception, bids them farewell and heads back towards the greenhouse. She grabs a Segway and starts to ride toward Greenhouse 13.

'I'm on my way, Zoë. Can you give me an update?' Tess asks.

Some of the greenhouses have broken. The storm is getting in and wrecking the crop. We need to decide if we can do anything about it, Zoë responds.

'What do you mean, broken?'

There are some huge hailstones in the storm. They're smashing the roof. Be careful.

Tess looks up at the roof again. It's hard to see what's hitting it, but the noise is apocalyptic.

Looking ahead, she can see rain falling through a large hole. She is still only in Greenhouse 11, so the damage must be widespread. She approaches cautiously and can see hailstones the size of handballs lying amongst the shattered glass. Some of the tomatoes have shards of glass embedded in their soft flesh.

'Holy mother of God,' she mutters. 'I'm glad I'm inside. If I was hit by one of those hailstones, it would have killed me.'

She diverts down one of the rows to avoid the hole in the roof, weaving her way around the robots that monitor, feed and harvest the plants.

'There's damage here in Greenhouse 11 as well,' Tess says, paging Zoë.

There is no response, so she tries Mila.

'Mila, can you hear me?'

She doesn't receive any response into her EarBud and starts to feel a panic rising.

She tries to remain calm but she keeps looking at the ceiling, expecting it to cave in and release the deluge upon her. She hears the sound of shattering glass regularly as other areas of the greenhouse succumb to the weather.

'Plan, plan, what's the plan?' she mutters to herself.

She turns the Segway around to head back towards the safety of reception. There is a deafening crash as a huge crack opens in the glass and Tess watches shards of glass start to fall. She is hemmed in on either side by the tomato plants and doesn't have anywhere to go.

She lets out a strangled scream as a large chunk of glass hurtles towards her and plunges through her soft flesh.

SARAH

CHAPTER 3

Later that day Ayesha, Cassidy and I set out looking for water. We haven't gone far before Cassidy complains of a funny tummy and returns to camp. Our path has been steadily downhill, but the scenery hasn't changed. There are plenty of pine trees and we find more of the large mushrooms, but no water. The ground under the pine needles is damp, suggesting it must rain regularly, but there are no lakes or pools collecting it and we haven't heard any streams or springs.

We pick some mushrooms and begin the climb back up the hill when I hear something rustle in the forest ahead and catch a glimpse of movement.

'Did you hear that?' I ask Ayesha.

'Hear what?' she replies.

I call out, wondering if one of the others has followed us, but no one calls back or jumps out of hiding.

On the walk back, I can't get the feeling of being watched out of my head, but I can't tell if it's a rational thought or just my imagination.

Back at the camp, Nat has constructed some tall, skinny walls made from thin branches and I spot Karim crouched in a hole in the ground. Declan is sitting with Tomohiro, watching proceedings.

After one more glance behind me, I walk over to Karim and sit on the edge, my feet dangling half a metre above the bottom of the hole.

'Careful, don't knock any dirt–' he starts to say, but I'm settled before he can finish. 'Too late,' he chuckles.

'What happened to Declan digging the pit?' I ask.

'He copped out after twenty minutes. Kept looking at me pathetically until I volunteered to help. Then, as soon as I joined him, he buggered off. Fair enough, though, this thing is bloody annoying to dig with.' He holds up a small, foldable spade. 'Luckily the ground is super soft.' His voice suddenly becomes quieter. 'Also, don't react, but I need to tell you something. In fact, laugh like I've said something funny.' He flashes me a winning smile.

I laugh, trying not to look too obvious.

'When you were gone, Declan and Nat tried to get me to join their alliance.' He continues to smile, like we're talking about nothing serious at all.

'What do they think this is?' I say. 'We're not on *Survivor*. It's not a game.' I try and keep my face nonchalant but talk in low tones so we can't be overheard.

'I know. They told me that they wanted me to vote with them. That if I did, they'd treat me well.'

'And you said?' I ask.

'I said I'd vote on each issue independently, based on what I think is right.'

'As we all should,' I say, trying to keep the exasperation from my face.

'Worst of all, they asked me to pretend to "side with" you and Cass. And to report to them what you guys talk about.' He grins at me and noticeably raises his voice. 'Well, that's very interesting.'

'How's the hole going?' Declan has appeared behind me.

'I want it to be deep enough. What do you think?' Karim replies.

'Needs to be deeper,' Declan says bluntly.

Karim turns back and continues shovelling. 'What were you saying about the mushrooms, Sair?' he asks casually.

'Oh, um. Yeah, that there's plenty to the southwest,' I reply, less smoothly. 'If Ayesha confirms they're safe, we'll have plenty to eat. Bit boring though.'

I excuse myself from the conversation and go to the hut to check on Cassidy, but she's not there. When I question Tomohiro, he says that she only came back to camp briefly before heading north. He tried to follow her, but she was gone before he could do anything about it. I head in the same direction, hoping that she hasn't gone far, otherwise I'll have to put together a bigger search party.

I pass the area where my fellow Eta volunteers are. Well, where their pods are, at least.

Kyle's once rectangular pod, now pinched in the middle, looks like a large, metal bowtie. The lid has popped open as the sides of the pod were slowly crushed by trees. It would've taken years and years, and yet Kyle didn't stand a chance. As soon as the seal was broken, the exposure to oxygen would've halted the chemogenic preservation process and started cell decomposition.

I steel myself and lift the lid to see a skeleton staring blankly into the canopy. I didn't get to know Kyle as well as some of the other Etas, but he was always smiling and had kind eyes. At his feet, I can see the remnants of his spare clothes and a broken pile of books. Most of them have decayed away, leaving only spines filled with holes. He was a lawyer and a teacher.

The food is also wasted. His water, stored in stainless steel canteens, is the only thing intact.

Taking the water from his pod, I close the lid and say a silent prayer, thanking him for his sacrifice.

I look at the tree that stands where Ang's pod should have been, right next to my own. If that tree had grown two metres to the left, my pod would've been the one to vanish.

I don't know why I deserve to live when so many others didn't make it, but it fills me with a determination to make the most of this second chance. Us survivors need to avenge the deaths of our friends and colleagues and bickering amongst one another isn't the way to do that. I resolve to try and get along with Declan. He can play games, but I'll focus on keeping the peace and making my lost comrades proud.

Straining my neck, I follow the trunk of Ang's tree up, through the branches to the canopy. Taking a few steps back, I catch a glint of metal at the very top. Ang's pod hangs suspended between two trees, lying on its side with the lid partially open. A skeletal foot hangs through the gap between pod and lid.

Turning away, I blink tears from my eyes.

'We'll make this right,' I say softly to them as I continue on my way. 'I promise.'

I plod northwards, reminiscing about training with Kyle and Ang. With Pam and Mabel. With Sam and Meaghan. Years and years ago we learnt together, practised together. And yet, to me, it feels like only days ago. I still find the sudden and massive shift incomprehensible. How can two smart, happy and very alive boys turn so suddenly into skeletons?

I'm so caught up in my own thoughts that Cassidy sees me before I see her. She's only ten metres away, standing over an open pod.

I re-orientate myself to the surroundings and realise we're at the Epsilon pods. Cassidy's chemo-pod is nearby, and she has excavated and opened one of the half-buried pods.

'Hi, Sarah,' she says, a bit sheepishly.

I walk closer and see a man lying in the open pod. He has an oxygen mask over his mouth and his body is covered in a mesh frame that pierces his body with thousands of tiny needles. In the centre of the frame, over his chest, sits a metal, fist shaped object.

'What are you doing, Cass?' I ask.

'I... Well. This is Newton,' she says, not answering the question.

'And what are you doing with Newton, Cass?'

She looks at me with guilt in her eyes, like a child who's been caught rifling through the pantry.

'I dug him out. I figured he hadn't reactivated because his sensors were blocked. But then I got impatient and I decided to start manual reactivation.'

I see now that it's not just guilt in her eyes, she's also nervous. Worried that she's made a mistake, that she's not doing the right thing.

'Are the needles active yet?' I ask.

'No, I started having second thoughts and I...' She trails off.

'It's too late for that now. You've opened his pod. The oxygen will either help wake him up, or it will rot his body where it lies.'

'I know. I just don't seem able to do it.'

'Why did you decide to wake him in the first place? That should've been a group decision,' I say.

'Do you want the truth, or the reason I'm telling myself?' she asks.

'Both, I guess.'

'After Tomohiro and Karim were both poisoned at dinner, I thought of Newton. He's a Zoologist and a Chemist, so I thought maybe he'd know more about what animals are safe and which are poisonous. And he can run basic tests to detect dangerous chemicals.'

I look to Newton's feet and see the equipment needed for a very simple lab. Some glassware, bottles of chemicals and packets filled with paper strips.

'And?' I ask, knowing there's more Cassidy has left unsaid.

'And once I thought of Newton, I realised I was feeling lonely. You have Karim and the Alpha pack have each other. I thought it'd be nice to have another Epsilon around.'

I understand. Even with Karim, I feel outnumbered by the Alphas. Still, it wasn't Cassidy's decision to make on her own.

'Of course we're gonna try and wake the others, Cass, but when we do, it'll be as a group. But seeing as you've come this far, time to bite the bullet.'

'Can you do it?' she asks, looking for support.

'This is on you, Cass.' I walk up to her and place my hand on her shoulder. Her hand is resting on the manual override lever. 'On three, okay? One, two, three.'

She pulls the lever down and a very slight whirring sound begins to emit from the chemo-pod. From the thousands of needles, a chemical substance diffuses into Newton's major blood vessels. The oxygen mask starts forcing oxygen into his lungs and the metal fist presses on his heart, forcing the sluggish blood around his body as it slowly loses viscosity. As the blood is pumped through the main arteries, it circulates the chemical along with oxygen to smaller and smaller vessels. Eventually, the chemical will reach every cell in his body, slowly starting the process of cellular re-energisation.

Cassidy holds my hand tightly as we wait. There's nothing more that we can do except hope the process runs smoothly. The very same process that occurred inside our own pods only twenty four hours earlier.

After a while, the mesh frame lifts up, retracting the needles from Newton's body and lifting the fist away from his chest until the whole contraption is nestled in the open lid. A protective cover slides out of the lid and over the needles. We can see Newton's chest rising and falling of its own accord and, placing a hand on his wrist, I feel a steady pulse. I nod to Cassidy.

Newton is an attractive ginger. Slender, tall and a bit nerdy looking, thanks in part to a pair of silver framed glasses that peek out from under an unruly orange fringe. It appears he's trying to be a trendy nerd. He wears the same uniform as the rest of us, but with his sleeves rolled up and the bottom of his pants cuffed, exposing his ankles.

We wait another five minutes before Newton's eyes flutter open and he looks up at us in confusion. He tries to sit up, but collapses back into the pod.

'Remember the training, Newton,' Cassidy says. 'Take it slow. Start with the toes, the fingers. There's no rush.'

He complies, warming up each body part before he tries to sit up once again, this time with greater success. He looks around the pine forest, nodding to himself.

'Long time, no see, Cass. So, we're here then?'

'Yup. Welcome to the future.'

'Not that it looks much like the future,' I add. 'More like the past.'

'This is Sarah, by the way. Sarah, this is Newton.'

'Nice to meet you,' I say.

'Yeah, I remember you,' he says, smiling. 'Hard to forget a pretty face like yours.'

I blush and have to turn away, pretending to check something on the pod. 'Does everything feel alright?' I ask.

I'm not usually so receptive to flirtation, but something about Newton's smile and the way he looked at me makes me feel that he really meant it. The fact that my face burned so quickly shows that, for some reason, I care what this man thinks, even though I've only just met him. God damn it!

'I feel fine. What about you guys?' he asks.

'Fine,' we say together.

'We woke up yesterday,' Cass says.

'Oh, right. Anyone else?'

'Yeah, there's a few of us. And more like you, whose pods haven't reactivated yet,' I say.

'Come meet the others,' Cassidy says, motioning towards the Alpha pods and our fledgling settlement.

We walk towards the others and pause on the edge of the clearing. Four open chemo-pods line one edge of the settlement and behind these, still in the trees, three more lie under a rotting log.

In the middle of the clearing stand the two huts that we've constructed under Natalie's tutelage, and a small fire surrounded by some log stools. On the far side, just beyond the tree line, I can see that Natalie has finished the toilet. A small wooden building peeks out from the gloom, and a sign hangs above the door, proudly labelling it *The Shit House*.

'How long did you guys say you've been awake?' Newton asks.

'Since yesterday morning,' I reply.

'Nice!' he says, nodding approval. 'You've been busy.'

Declan and Natalie are lounging on the ground near the fire, Declan's arm casually draped over Nat's shoulders. Karim sits opposite them with Tomohiro, deep in conversation. Ayesha is noticeably absent.

Though critical of Cassidy's decision to wake Newton on her own, I want to protect her from the others, so I hold her hand as we approach the fire. She gives my hand a small squeeze.

Tomohiro looks up as we approach. 'Look who decided to come back,' he says, his voice slurry. His gaze shifts to Newton. 'And who's this, then?'

Declan and Nat glance over their shoulders and turn around.

'Newton. He's gonna keep us alive,' I say.

'And how exactly can you help, Newton?' Declan asks, his brow furrowed.

'He's going to keep us from being poisoned,' Cass says.

'Sounds good to me,' Tomohiro says with a lopsided smile. 'Although, you're a bit late.'

'Did you wake him up?' Declan asks.

Cassidy nods.

'That wasn't your decision to–' Nat starts to say before I cut her off.

'If it's a decision that will increase productivity by keeping us all safe, then it's the right decision,' I say firmly.

Karim and Tomohiro nod and Declan is smart enough to let this one go. Though he scowls at me as if it was my plan all along.

'Um, guys!' we hear Ayesha call out from the far side of the clearing.

'Yeah, are you alright?' I ask.

'Fine. Just, um, there's no toilet paper.'

We all burst into laughter.

'Just let it dry,' Karim calls out helpfully.

'That's not gonna work. It's, well, it's a code brown.'

The boys snigger. I turn to Cassidy.

'I guess we'll have to use the cloths and then wash them?' she says.

Nat wrinkles her nose.

'We could use moss,' Declan says.

'Have you seen much moss around?' I ask. He shakes his head.

'Guys!' Ayesha calls.

'Yeah, we're just discussing, Ayesha,' Nat calls back. 'You'll have to wait a minute.'

'Okay!'

'If we had running water we could just wash, but we don't, so cloth it'll have to be, I guess,' I say, looking around the group.

Everyone's face shows varying levels of disgust, but they all nod agreement.

'But we all hang on to our own stuff, yeah?' Tomohiro asks through droopy lips. 'I'm not washing anyone else's dirty… yuck.'

We all murmur agreement. Cassidy goes to her things and cuts up small squares of cotton. She delivers a small pile to Ayesha, and then hands out a quota to each of us.

'You lose yours, that's your problem. We can't use more, especially if we need these for the others that are yet to wake up,' she says holding her cloth squares in the air.

'We really need to find water,' Karim says. 'Any luck, Cass?'

'I got distracted waking up Newton. Sorry.'

'Then we all need to go looking tomorrow,' he continues. 'We have a good supply, but that will run out quick smart, especially if we have to use it for washing too.'

Ayesha re-joins the group, smiling at us a bit sheepishly.

'I agree,' I say.

'Tomorrow morning, we split up and each head away from the camp in different directions looking for water,' Declan says.

'No way,' I rebut. 'We need to at least be in pairs. I think I saw something out there.'

'Sure you did.' Declan says.

'And either way, we don't know what's in the forest,' I continue, ignoring his interruption. 'Until we do, it's safer in pairs.'

'We'll cover more ground on our own,' Nat says.

'You saw the teeth marks in that mushroom. If there's something here, it's big,' Cassidy says.

'If they even were teeth marks,' Declan says.

'What are you guys talking about?' Newton asks.

'Cass and I found a mushroom that was half eaten,' I say. 'There were grooves that we think are teeth marks.'

'Can I see it?' he asks.

I look around the group, wondering where the dried mushroom ended up.

'Sorry, mate. I used it to make the fire,' Tomohiro says apologetically. 'But it burned really well.'

'*If*,' Declan says, putting too much emphasis on the word, 'that mushy was eaten by an animal, it's probably moved on by now. We haven't seen any other signs of it.'

'Except that I just said that I think I saw something today,' I say, exasperated.

'Minds play tricks when you're in new places, Sarah,' Tomohiro says, in a surprisingly patronising tone.

'Well, why don't we vote on it?' Cass asks.

Most people nod. Declan looks around the circle, judging numbers and then agrees.

'In favour of splitting up to cover more ground looking for water?' Tomohiro asks, his words slurred.

Tomohiro, Natalie and Declan raise their hands. Nat glares at Ayesha who hesitates before raising her own hand. Declan stares down Karim, but Karim just smiles back broadly.

'Four votes,' Tomohiro says. 'Those in favour of pairing up?'

Cass, Karim, Newton and I raise our hands.

'Four votes. I guess that means no one is abstaining. We have a tie.'

'Which reminds me, I've been thinking we need to nominate a leader to help with decisions. And split ties,' Declan says.

'And who do you suppose that should be?' Karim asks sarcastically.

'Well, actually, I'd like to nominate Tomohiro,' Declan says.

'What do you say, Tomohiro?' Ayesha asks.

'Nah, mate, too much responsibility.'

'I nominate Declan,' Nat says.

'Surprise, surprise,' I mutter.

'I accept,' Declan says. 'Any other nominations?'

'I nominate Sarah,' Cassidy says.

'But you're a doctor, Cassidy,' Declan says in surprise. 'Why would you nominate Sarah? She's just a nurse.'

'Fuck you, Declan.' Cass says. 'I nominate her because she'll make a good leader and treat everyone fairly. And because she's a nice person. That's why.'

'Do you accept the nomination, Sarah?' Newton asks.

I pause for a second, remembering the dead volunteers and my promise to them to avoid bickering. But this isn't bickering, this is a chance to forge a better society.

'I accept,' I say. 'But I still think most decisions should be made by the group. Especially while we're still so small. If we have a leader, they should only decide on tied votes.'

'Or if we're separated and can't all vote,' Ayesha says.

'Or if an instant decision is needed,' adds Declan.

'Sounds good to me,' Newton says.

'Me too,' says Tomohiro.

'Any other nominations?' Cassidy asks.

Everyone is quiet.

'Then we vote. Sarah or Declan,' Nat says.

'Hang on,' Karim says. 'Shouldn't we vote confidentially?'

'Too much hassle,' Declan replies.

'We will all find out who voted for who anyway,' I add. 'But good thought, Karim.'

'When do we reconsider the leader?' Ayesha asks. 'I mean, do we have yearly elections. Can anyone contest at any point?'

'Also good points,' I say.

'At the very least we should do a re-vote a week after we wake everyone else up. Once everyone has had a chance to get to know each other,' Tomohiro says.

'And then should we have an "election" every six months,' Cassidy says, using air quotes.

Everyone agrees.

'One more thing, though,' Ayesha adds. I'm surprised by her confidence. Normally one to sit back, she's very engaged with this 'political' conversation. 'A new election should be able to be called if two thirds of us want it, regardless of how soon it is since the last election.'

'What did you teach again, Ayesha?' I ask.

'Politics and History,' she says with a sheepish grin. 'And Botany of course.'

'If that's all sorted, we should move to voting,' Natalie says. 'Votes for Declan.'

The four Alphas raise their hands.

'Votes for Sarah.'

The others raise their hands alongside my own.

'And no one abstains. We have another tie. Surprised, I am not,' Nat says derisively.

'Then we campaign,' Declan says.

'Good idea,' Newton adds. 'We can vote again same time tomorrow, and Sarah and Declan have until then to try and change our minds.'

Cassidy leans forward. 'But that doesn't help with our discussion about searching for water tomorrow. Can we go in pairs?'

'How about this,' Tomohiro says. 'Those of you who voted to go in pairs can go in pairs if it makes you feel safer. The rest can go separately to cover more ground. And I'll just stay at camp, doing nothing.'

'We'll see about that,' Karim says. 'I'm sure we can find something for you to do.'

'Is that everything decided then?' I ask. 'Because if so, I'd love some grub. What's for dinner?'

SANTIAGO

Sabastián looks across at Santiago and gives him that lopsided grin he always uses when he's about to make a wisecrack.

'Ya gonna keep working, boy? It's not that hot that ya can't carry a bunch o' bananas is it?'

To be fair, it is too hot. The radiation coming off the sun feels as if it could burn your skin in minutes and the air is thick with humidity. Sabastián, Santiago and the other harvesters are all topless and covered in sweat. None of them should be working, but the crops need to be harvested. People need food to survive.

Each year, as temperatures have continued to rise, the harvest has become progressively harder. But this year, with daytime temperatures bordering on sixty degrees Celsius, the harvest has been hellish. To avoid the worst of the heat, they sleep in the daytime and harvest at night. They were given a break from harvesting the last few nights as the mercury stayed dangerously high, but with supplies running low, the decision was made to recommence the harvest last night.

Santiago and the others work hard to complete as much as possible before sunrise, but with overnight temperatures offering little relief, they didn't reach their quotas and need to continue into the morning.

Sabastián, a veteran of the harvest, is just trying to keep spirits high, but Santiago is tired. He has been hauling large bundles of bananas for hours and now that the sun is on his back, he is beginning to dangerously overheat.

Santiago can feel his breath getting shorter and shorter, but he knows the bananas need moving. Pushing off a wave of nausea, he gulps down some water and heaves a bundle of bananas onto his back.

'Oi. Where're ya off to?' Sabastián calls out, as Santiago starts lumbering away.

'The packing shed,' he replies.

'Then ya better turn around. Ya walking into the plantation.'

'Right. Yeah, of course,' he mumbles.

Santiago tries to turn around, but he can't remember which way he was facing to begin with. He is hot. Too hot, and why isn't he sweating anymore?

He drops the bananas and stumbles around, looking for the water bottle.

'Oi, watch the produce,' he hears someone cry out.

Santiago tries to blink the stars from his eyes, but they won't go away. Another wave of nausea flows through his body and this time he hurls all over the grass. Santiago falls to his hands and knees, landing right in the spew pile and causing more vomit to spill from his body.

Sabastián rushes over to help. He picks up the water bottle and pours it over Santiago's head and shoulders, but the boy topples sideways to the ground. Santiago's breathing is fast and shallow.

Others arrive to try and assist, all sweating furiously themselves and fighting against the oppressive heat, but Santiago cannot be saved. His body begins to spasm and he cries out in pain.

Later, those present can't decide what was more disturbing. The jerking, crying, wheezing and vomiting. Or how all the noise and movement suddenly stopped, as Santiago's mind and body finally succumbed to the heat.

SARAH

CHAPTER 4

The next morning, after a tasteless breakfast, we prepare to search for fresh water. We gather water canteens, fungus and dehydrated food into our rucksacks and discuss a plan.

Predictably, Declan has plenty to say on the matter. 'I think we should each strike out in different directions and after a certain distance, we turn and walk towards where someone else should've made it too, though that person will have already moved on. That way we can scout a circle around the camp and intersect any streams.'

I think I understand what he is explaining, and I'm pretty sure it's a bad idea.

'Remember some of us are going in pairs,' Cassidy chimes in. 'I'm with Newton.'

'And I'm with Sair,' Karim says, adjusting the straps of his backpack.

'Which directions would we go?' Nat asks.

'With five groups, we divide the compass evenly–'

'That's a group every seventy two degrees,' Karim says, cutting Declan off. His mental maths is far superior to my own.

'If I go with Nat, then we're four groups. That divides better, yeah,' Ayesha says nervously. Everyone is silent for a moment, understanding Ayesha's back track on her vote with the Alphas the night before. Declan scowls at her. I file it away, making a mental note to try and persuade her to vote for me. Maybe she's the weak link.

'Four groups do make more sense,' Karim says. 'Ninety degrees. One point of the compass each.'

'Yeah, alright,' Declan says begrudgingly. 'But let's do north-east, north-west, south-west and south-east.'

We all agree, giving Declan final say in the matter. While they chat, I try to pinpoint my issues with the plan.

'And how far are you going?' Tomohiro asks. He is slouched against a stump and looking gloomy at the idea of being left alone at camp. He can crawl to the Shit House if he needs to but is otherwise dependent on others for movement.

'Ten kilometres,' Declan says confidently.

'If I understand your plan correctly, Dec, then that's thirty four k's each. Too far,' Karim states.

'Really?' Declan asks.

'Uh huh. Ten k's out, fourteen across to the next point and ten back.'

I marvel again at his mental speed, even managing Pythagoras without a calculator.

'How far in total if we only went out five kilometres?' Ayesha asks.

'Seventeen,' he replies instantly.

'Seven?' Nat queries.

He takes a bit longer, bobbing his head from side to side as he thinks. 'A tick under twenty four.'

'At six k's an hour, we'll be back in four hours. Sound good?' Declan says, more a statement than a question.

'Sorry to be a downer, but this plan sucks,' I say.

Declan glares at me, but I push on.

'The terrain around here won't be flat, and we don't have an accurate way to measure distances. If we are off by a few degrees in our navigation, we could so easily get lost.'

Karim is nodding. 'We could easily end up hundreds of metres off course.'

'No thanks,' Nat says.

'Fine, in that case we just walk seven k's out and back,' Declan says retaking control. 'Should only take three hours.'

'I don't walk that fast,' Ayesha says.

'Even at four kilometres an hour, you guys will be back in six hours. You'll be fine,' I say, trying to reassure her.

She gives me a weak smile.

We decide on which direction each group will go and complete final checks.

Nat gives Tomohiro a plan for some wood that he can saw into smaller pieces using his good side. But she won't tell him what they are for. This piques my interest and I look over his shoulder at the drawing. She wants lots of long planks and a large log sliced into four rounds.

Realising that I'm alone with Tomohiro, I try to strike up a conversation, but he cuts me short, pre-empting my intentions.

'Not gonna happen, Sair. I'm weak now. That means I have to follow power.'

'But what makes Declan powerful?' I ask. 'What does he have over you?'

He looks at me with compassion. 'You'd make a fine leader Sarah, but you don't need to lead. You don't aspire to it. Declan hungers for it. Without it, he thinks he's nothing. That hunger makes him ruthless. That ruthlessness and unrelenting will, that's his power.'

I shake my head, knowing there's some truth to what he says. Declan will stop at nothing to take control. I'll respect his leadership if he wins. Would he respect mine? I doubt it.

I spot Ayesha taking a last-minute toilet stop and intercept her once she's done.

'Can we chat?' I ask, as she steps through the fabric flap that acts as a door.

She nods and we take a few steps deeper into the trees.

'Who are you gonna vote for tonight?' I ask bluntly. She doesn't meet my gaze so I change tact. 'Why Declan?'

'I want to vote for you Sarah, I really do.'

'Then why don't you?' Something just doesn't sit right with the way the Alphas continue to follow Declan. Nat appears smitten, so I've written off trying to change her vote, but Ayesha is sensible. She doesn't follow Declan blindly, but she continues to follow him. Why?

'I just can't,' she says.

'Are you scared of him?' I ask.

She nods. She looks up at me briefly and I smile to reassure her, but she looks away.

'What if I can promise to protect you?' I say.

'I'm not sure you can,' she replies. 'He's really smart and such a snake. He'll find a way.'

'At least promise me you'll think about it? If the vote is tied again, nothing happens. But if you vote for me, I'll lead us with strength, but also compassion. Cassidy, Newton, Karim and I will protect you from whatever it is Declan might do.'

She nods. 'Like I said, Sair. I want to vote for you.'

'Then you know what? I think that you might just do it.' She smiles at me. 'Think about it, yeah?' I squeeze her hand and walk back to the clearing.

I join Karim and smile broadly at him as we head off, following our compass north-west. We're unable to talk much, both counting steps in our head to estimate the distance covered.

'One thousand!' I say triumphantly, marking the first kilometre complete.

‘I’m only on eight-hundred-and-thirty-six,’ Karim says with a smile.

‘Your legs are too bloody long,’ I say playfully.

‘Excuse me, miss, but maybe your legs are too short.’

We agree to call nine-hundred ‘Karim steps’ a kilometre.

The walk is steadily uphill and at times we have to clamber up small cliffs. It makes a nice change to see some large rocks protruding from the dark soil, but at the top of each rise the pine forest takes over once more.

Occasionally, I forget how many steps I’m on, but Karim doesn’t miss a beat, his mind working like a computer. We let the silence sit comfortably between us as we count steps in our heads, eyes scanning the surrounding trees, falling into the rhythm of the hike.

As we approach the seven-kilometre mark, I start to notice the faint sound of running water.

‘Do you hear that?’ I ask.

‘Hear what?’ Karim asks.

‘I think I hear water. Let’s keep walking.’

The sound gets louder, almost imperceptibly, but I become surer that the soft gurgle I hear is what we’re searching for.

‘Nine hundred,’ Karim says, breaking my intense concentration on the sounds around me. ‘That’s seven k's.’

‘You really don’t hear that?’ I ask Karim.

‘Hear what?’ he asks again, continuing to score a tree.

‘Stop and listen, Karim. That gurgling sound.’

He tilts his head to the side. ‘Yeah, maybe.’

He may he better at mental maths than me, but clearly my hearing is superior.

‘How’d you even get through the screening process with ears like that?’ I ask jokingly. ‘Come on, we have to check it out,’ I say.

I jog off ahead of him and he calls out to me as he tries to catch up. ‘I think it must have been my good looks.’

The route continues uphill and I'm panting and sweaty by the time I find the source of the gurgling noise. Karim arrives soon after.

I spare a smile for him. 'It was a rhetorical question, dipshit.'

We both look down at the small stream and smile. In survival terms, water means life.

The stream flows languidly, but the water looks clear.

'You game?' Karim asks.

In answer, I bend over, cup my hands together and drink a mouthful of water.

'Delicious,' I say flicking the excess water off my hands at Karim.

We follow the stream a few hundred metres uphill to where it emerges from a flat bed of rock. The spring trickles forth from a large crack in the rock, makes a small pool near the bottom, and overflows to begin the stream. Unfortunately, the general direction that the stream flows is north-west, down the opposite side of the hill and away from our camp.

'Seven k's just to get water. That's a hike,' Karim says.

'Let's hope the others have found something closer,' I say.

'But hey, freshwater is freshwater. Fucking ay?'

'Fucking ay,' I reply.

I judge that we are approximately at the summit of the hill we've been climbing, but even up here, the pine trees grow tall, obscuring the view and making it hard to tell.

We fill our canteens with spring water and head back to the seven-kilometre mark of our hike. The freshwater tastes about a million times better than the stale, thousands-of-years-old water we've been drinking.

'Back to camp then,' Karim says.

'Yes, sir.'

He checks the time on his solar powered watch. 'We should probably have some lunch first, then we can head straight back,' he says.

'I wish we'd brought more stuff with us. More supplies, more solar-powered shit.' Something about being calm and relaxed around Karim brings out my potty mouth.

'I know, right, but I guess we couldn't bring everything.'

After a quick lunch of raw mushroom, we head for home and without needing to count steps, we chat idly.

Now that I'm not concentrating so much on where we are going, I also become more aware of my body. My feet are sore from the seventeen-odd kilometres we've hiked, and my neck and shoulders are tight, most likely from sleeping on the forest floor.

Once we are nearly back at camp, we hear movement ahead, and I'm surprised to see Cassidy storming towards us. Newton hurries to follow her.

'What's the matter, Cass?' I ask.

'See for yourself,' she says, walking past.

'Cass, wait up,' Newton calls. 'In there,' he says to us, gesturing over his shoulder. 'I'll calm Cassidy down.'

We enter the camp clearing and I see Declan talking to someone that I assume to be Tomohiro. He hears us approaching and turning to see us, he smiles. A look of smug triumph fills his eyes.

He shifts slightly, revealing the person he sits with.

'Karim, Sarah. I'd like you to meet Jay,' Declan says. He looks me straight in the face and winks. *Anything you can do, I can do better.*

'Hey, Jay, nice to meet you,' Karim says diplomatically.

Jay sits on a stump in the same utilitarian uniform as the rest of us. He has dark skin, dark hair and a broad smile. I can't tell if he is yet aware of the fact that he's just another piece in Declan's game.

'A pleasure to meet you both,' he says enthusiastically. 'What a world, hey? Can you guys believe it? I guess maybe you can, you've had a bit more time to take it in, but wow! And it's lovely to meet you both properly, it really is.' A talker then.

'Nice to meet you too,' I say, shaking his proffered hand. I turn back to Declan. 'And what support can Jay provide us?'

'He's a Physio. He can help Tomohiro recover,' he replies plainly.

I nod, knowing full well that's not why Declan has woken him up.

'Plus, my Aunties and Uncles have taught me heaps about bushcraft since I was a little tacker,' Jay adds. 'Not that there is much bush around anymore.'

'Let's go check on Cass,' Karim says, pulling me away.

'I'm also a butcher,' Jay calls after us. 'So I can help prepare any animals we catch. But it was nice to meet you both. See you soon, yeah?'

I wave goodbye over my shoulder and head back into the forest with Karim.

We follow the sound of Cassidy and Newton's voices and find them in a small clearing. Cass is slumped on the ground and Newton is crouching next to her.

'You alright, Cass?' I ask.

'He's just such a twat!' she replies. 'He's done this just so that he can win the vote.'

'Who is Jay, anyway?' Karim asks. 'Gamma?'

'Alpha,' Newton replies. 'He was one of the pods under that large log. Declan spent the afternoon waking him up rather than hiking to look for water.'

'What a turd,' I say. 'But I guess you set the precedent waking Newton, Cass.'

'I know. That's why I'm so mad. It's kind of my fault.'

Karim shakes his head. 'No, it's not. Without Newton, we would've lost the vote yesterday anyway.'

'That's true,' Cassidy says. 'He tried to get us to side with him this morning.' Newton nods beside her. 'When we said no, he threatened the other Epsilons.'

'What did he say?' I ask.

'Nothing direct,' Newton says. 'He just talked about the fragility of the pods and how vulnerable the slumbering volunteers are. We knew what he meant.'

'He wouldn't, would he?' Karim asks.

We're all silent, none of us knowing the answer.

'Even if Declan has Jay, we may still have the numbers,' I say.

They all turn to look at me, surprised and hopeful.

'Tomohiro?' Karim asks.

I shake my head. 'Ayesha. I think I got through to her. If she votes with us, I win five to four.'

'Let's hope she does,' Newton says.

'Best news I've heard all day,' Cassidy says with a smile.

We head back to camp and see Jay and Declan carrying Tomohiro to the fire area. Soon after, Ayesha and Nat emerge from the forest. Ayesha rushes forward and wraps Jay in an enormous hug. Nat follows behind, more circumspect. She moves to stand behind Declan, and he gives her a quick smile.

Cass, Karim, Newton and I go to sit down opposite the others but, thinking better, I insert myself between Declan and Jay, not wanting to show obvious division.

'So, is this everyone then?' Jay asks. 'What a group. I mean it's too bad there's not more, but you've done alright, haven't you?' He gestures to the camp around us.

'It's good to have you, Jay,' Ayesha says with a huge smile.

'Any luck, you too?' I ask Ayesha and Natalie.

Nat shakes her head.

Ayesha opens her rucksack and pulls out some small mushrooms. 'No water, but we did find a new type of mushy.'

'Newton. Cass. What about you guys?' Karim asks.

They both shake their heads.

'Mostly just pine trees on our route,' Cassidy says.

We all turn to look at Declan.

'I'm guessing you didn't find anything,' I say.

'Well, I didn't get very far,' he replies. 'I was walking through the forest and all I could think about was poor Tomohiro, alone at camp. And then I thought of Jay and I knew I had to wake him up to help out our old mate here.' He gives Tomohiro a tender smile, playing the part of a caring friend with conviction.

'That's the sweetest thing I've heard since waking up,' Nat says, playing her role.

'Well, shit,' I say, dispirited.

'You didn't find anything either?' Ayesha asks.

'No, we did actually,' Karim says.

I shuffle on my log to get more comfortable. 'Only problem is that it was beyond our original seven-kilometre hike. So, our closest drinking water is more than seven clicks out from camp.'

'What was it?' Cassidy asks.

'A mountain spring,' I reply. 'It emerged near the top of a slope and ran as a small stream down the other side of the hill, away from camp. It's good water though.'

Karim and I pass our canteens around and everyone tries the spring water. They all smile when the fresh water touches their lips.

'Well, we know where it is,' Newton says stoically. 'That's the main thing. It may be a hike to get there, but we've got water.'

Everyone nods agreement.

'Maybe we should move camp up there?' Tomohiro suggests. I notice that his words seem a bit clearer than yesterday. Hopefully the paralysis in his face is lessening, but it's more likely that he is just getting accustomed to talking with his paralysed features.

I ponder the idea, but Nat is quick to shut it down. 'We've already built huts here,' she says. 'Would we really want to start again?'

'Plus, I wanna stay close to the chemo-pods,' Karim says. 'I wanna try and rig something up.'

I put the matter to rest with one final observation. 'It was also really rocky and uneven up there. I'm not sure there were many places suitable for buildings.'

'That's sorted then,' Declan says.

'Still, it was worth thinking about,' Tomohiro grumbles.

Cassidy smiles at him supportively and pats his hand.

'Back to Declan and Jay,' I say, changing the topic. 'I think we need to have a serious discussion about reactivating the remaining pods. First Cass, then Declan. I understand your thinking, but these should be serious group decisions, not individuals off on a whim.' No one answers me, so I continue. 'When are we going to wake the others? What's the plan moving forward?'

'How many others are there?' Jay asks. 'Any more Alphas?'

Declan shakes his head. 'The tree that fell on your pod has burst both of the chemo-pods next to yours.'

'There's two more Eta pods,' I say. 'They looked to be intact, although they're both on their sides.'

'And two pods half-buried in the hill. One from Epsilon and another from Gamma,' Cassidy adds. 'The rest of the pods were covered in the landslide, so we have no idea what they're like.'

'So hopefully four more, but we don't know for sure,' Karim summarises. 'And an undetermined amount in the hill, up to a possible twelve more.'

'More people means more hands to work,' Newton says.

'But also, more mouths to feed,' Declan replies.

'Well, what are you thinking Sair?' Karim asks. 'Should we wait or wake them?'

Everyone turns to look at me and I have a sudden understanding of what being the leader of this ragtag group would be like. Even Declan seems keen to hear my thoughts and I take my time before answering.

'I think the knowledge, skills and manpower that they'll provide is invaluable, but I also think it's our duty as the first wave to make sure it's safe before bringing more people into this world.'

Most of the group nod in agreement.

'I agree with Sarah,' Declan says, respecting my opinion and thereby making this matter irrelevant in the upcoming vote. 'We should build more buildings and look for more food and water. Maybe in a week we'll be better placed to wake the others.'

'Now that's all settled, should we move on to the leadership vote?' Nat asks. 'Does everyone know who they're going to vote for?'

Everyone gives a sign of agreement. I glance at Ayesha, who gives me a faint smile before looking away. Jay nods along with the rest, indicating that Declan has filled him in on the leadership contest, and that his vote is secured. I consider delaying the vote to talk to him individually but, trusting in Ayesha, I decide to forge onwards.

'Let's start with votes for Sarah then,' Nat says.

I scan the circle, seeing the expected raised hands of Karim, Cassidy and Newton. I look to Ayesha, hoping to see her hand in the air, but both hands remain in her lap. She shrugs at me apologetically, glances at Declan and mouths, '*Snake.*'

I was hoping for a surprise victory. For Ayesha to show faith in me and in doing so, outmanoeuvre Declan's ploy of waking Jay. But I realise now that he secured Ayesha's vote in that same move. He confirmed her suspicions that he's in control and that we couldn't keep her safe. Ultimately, her fear of Declan won out.

'And all those in favour of installing Declan as leader,' Nat continues.

Five Alpha hands rise into the air in unison.

'Congratulations, Declan,' I say, forcing a smile onto my face.

'Long live the king!' Tomohiro calls.

'Long live the king!' Jay echoes.

'Thank you all so much,' Declan says with false modesty. 'I don't deserve to lead such an amazing group of young people, but I will do my best to be just, strong and true.'

It's too much. I stand up and walk away before I have to vomit.

Someone follows me away from the campsite and when I turn, I'm surprised to see Newton smiling at me.

'Do you like to climb?' he asks me.

'Sorry, what?' I say, caught off guard.

'Rocks, trees. Do you like to climb?'

'Um, I don't really know. I haven't done much before.'

'Let me teach you, then.'

We walk deeper into the trees. I notice that my hands have gone sweaty. Not a good start to my climbing career, or to being one-on-one with a boy.

'Did you get a view from up top?' Newton asks. 'I mean, over the valley?'

I shake my head. 'Pine trees everywhere. I haven't glimpsed anything outside this forest. It seems damn huge.'

He nods agreement. 'Let's see if we can change that. This one will do.'

He stops at the foot of a large tree, its trunk thicker than those of its neighbours.

'The trickiest bit is the first branch,' he says, pointing to a branch about two metres up and out of my reach. 'I'll give you a boost. After that, it's just like climbing stairs.'

I try to wipe my hands on my pants, but they just slide over the waterproof fabric.

'Are you sure about this?' I ask Newton.

'You'll be fine,' he says, cupping his hands together and motioning for me to put my foot in the artificial stirrup.

'Okay, here goes,' I say, trying to sound more confident than I feel.

I place my foot in his hands and he propels me upwards. My arms scrabble against the trunk as I try to keep my balance. The branch Newton pointed out is suddenly right in front of me, so I grab hold of it tightly.

'And up,' Newton says, giving my feet another shove.

I pull my belly onto the branch and then sit up, straddling the branch.

'Piece of cake,' Newton calls up to me encouragingly.

'Thanks.' I smile down at him. 'Now what?'

'Stand up carefully. You've got other branches you can hold on to, and then start climbing up, spiralling around the tree.'

I follow his instructions, climbing to higher branches and ducking through the narrower gaps. I rise another metre.

I smile down at Newton again. 'But how are you gonna get up?'

He smiles at me and then launches himself into the tree. He grabs the lowest branch and hauls himself up onto it, his biceps bulging through his sleeves as he climbs.

'Piece of cake,' he says again, panting slightly.

We ascend the tree together, gaining small scratches on our hands and wrists for our efforts.

Two-thirds of the way up, Newton spots a solid branch and gestures for me to sit. He sits close to me, keeping our weight close to the centre of the tree.

'Thanks for this,' I say. 'For the distraction, I mean.'

'No worries. Are you pissed Declan won?' he asks.

'Of course. He's just so arrogant. And it's just, I've never been a leader before. Not at girl guides, not in sport. And then here, I had a chance to be, and I thought about it, and you know what? I actually think I might've been good at it.'

'Of course you would've,' he says, turning to look at me. 'I didn't vote for you because you're not Declan, I voted for you because you're Sarah.'

In that moment, a life flashes before my eyes. A life in this new world where we survive whatever's thrown at us, where Newton and I have children and we watch them grow up together.

I lean towards him, hoping that he feels something similar, but he turns his head away and looks out into the forest.

'Maybe it's for the best?' I say, my thoughts returning to Declan.

'What do you mean?' Newton says.

'It's just something Tomohiro said to me that keeps going around in my head. *Declan needs to lead.* Even if I'd won the vote, I don't think I would've felt in control. I would've always been glancing over my shoulder, expecting Declan to make a move, or to try to undermine me.'

He nods his agreement. 'He seems to have a bit of a complex, doesn't he?'

'That's one way of putting it.'

'Should we climb to the top?' he asks. 'See what we can see?'

'Is it safe? The branches are getting pretty skinny.'

'It should be alright. Just check each branch gently before putting your full weight on it,' he says, his smile more reassuring than his words.

I make the mistake of looking down and seeing the five metres that I'd fall if a branch did snap on me. I push away the vertigo and steel myself.

We slowly climb higher into the tree. As we near the top, the central trunk starts swaying with our weight.

I feel queasy to the pit of my stomach. 'I think we should stop.'

Newton smiles at me. Oh, that smile. 'If we climb together, on opposite sides of the trunk, it'll stop most of the swaying,' he says. 'And look, we're almost above the other trees.' He points to the neighbouring pines, their tops just above our heads.

We climb together and a few branches later, our heads emerge above the canopy. I gasp in awe as I take in the landscape in front of me.

As far as I can see, from horizon to horizon, is the dark green of pine trees. Many, many kilometres in the distance I see a dark fissure running through the green, possibly where a river works its way through the trees.

As we're in the low mountains, the sea of green slopes away from us, but the mass of trees obscures all other land features.

The sky is brilliant blue overhead and the bright sun is harsh on my eyes. It is less humid in the open and a slight breeze kisses my skin. Angry clouds fill the sky far off in the north. Reds mingling with purples and grey. A flash of lightning fills one of the clouds and I point it out to Newton.

'Beautiful, isn't it?' he says, staring at the clouds.

We stand in the tree together, leaning against the trunk and looking out over our new world. I expect the sound of thunder to reach us, but none arrives.

'Should we be worried about that storm?' I ask. Another brilliant flash of lightning fills the clouds. I'm amazed at being able to watch such a raw, untamed thing, with a clear blue sky overhead.

'It's a long, long way away,' Newton replies. 'Can't even hear it. I think we'll be alright.'

I turn to the south, looking for a glint of water. We can't see the ocean from our vantage point, but we know that it's out there, over the top of the forest. Or at least it was, thousands of years ago. Looking at how the pines have spread, I realise that we can't be sure that anything is the way that it was before.

We watch the storm together, making noises of wonder and enjoyment. Even though we're both transfixed, we keep pointing out the bigger flashes to each other, as if somehow the other may have missed them.

Eventually, the gaps between the lightning flashes get longer and longer and we agree to climb back down.

'I forgot to mention that going down is actually a lot harder than climbing up,' Newton says with a cheeky smile.

'You dick,' I say, slapping his shoulder and almost losing my balance.

I very carefully make my way down the tree, lowering myself from branch to branch. On a few occasions, I have to hug the trunk as my feet search for purchase, and I get abrasions on both wrists from the rough bark.

Once I'm safely back on the ground, I show my wrists to Newton, who shows little sympathy. He holds out his own scratched wrists to show me.

'That's part of the deal, unfortunately. But it was worth it though, wasn't it?' he asks.

It's impossible to argue otherwise.

SEMYANG

'Mama, do you think we should be worried about the flooding?' Tshering asks.

He is an inquisitive boy and at thirteen, he is already more intelligent than Semyang ever was. She couldn't be prouder of her son, so she ensures she responds thoughtfully, while trying to allay his fears. 'I don't think so, darling. We're a long way downstream from the impacted areas.'

They live on the banks of the Puna Tsang Chu river in the Bhutanese mountains. In better weather, the mountains and trees make for breathtaking scenery.

'But isn't that how rivers work? It all flows down eventually?' he asks stubbornly.

'Of course, but it will hopefully dissipate a bit by then.'

'Did you know that one fifth of the world's population relies on water from the Hindu Kush Himalayan region?' Tshering asks.

'No, I didn't,' she replies, giving him an encouraging smile.

'But the glaciers have been melting for years. So, in the wet season we get dangerous deluges that no one needs and then there is drought in the dry season.'

'My son. I do believe that one day you may save the world. But you can't worry about everyone else all the time. Are you ready for archery practice?'

Later that night, when they're walking home from the archery range, Tshering listens to the radio on his phone. The usual program has been interrupted for an update on the flooding.

'Hey, Mama. They're saying that the dam has collapsed at Thorthormi. It couldn't stand the combined pressure of the glacial melt and the storms.'

Semyang looks down at her hands, as if trying to find an answer there. 'They've been warning about it for years. The more the glaciers melt, the more swollen the lake becomes.'

'Are we going to be alright?' Tshering asks.

'Of course, my love. We've nothing to worry about.' Semyang wishes that she felt as confident as her words.

They wake the following morning to a deluge of rain. The monsoon has hit with full force and even after the sun should have risen, it remains dark outside. They turn on the news and are devasted by what they hear.

We have reports that the entire town of Punakha has been washed away by flood waters, originating from the Thorthormi lake, the dam collapse, and continued downpours across the region. There are grave fears for all six-thousand residents, who may have been caught unawares overnight. In addition, it has been confirmed that the seventeenth-century palace of Punakha Dzong has been completely destroyed by the flood waters.

Authorities are warning all communities along the Puna Tsang Chu river to evacuate now. The government of Bangladesh has also released an emergency warning as they prepare for mass flood events downstream. The capital Dhaka is likely to be affected, and it is unclear how many of its eight-point-nine-million residents will need to be evacuated.

Elsewhere in Bhutan, landslides have killed at least thirty seven people as rain continues to fall on areas that have been made vulnerable by the ongoing glacial melt.

'We need to leave,' Semyang says, looking around the hut at their meagre belongings.

'But how? Where will we go?' Tshering asks.

'We will have to go on foot. We have to get away from the river. We walk uphill and try to find shelter.'

'It's pouring out there, Mama,' Tshering complains.

'I know, but you of all people should know. It's not safe here.'

'I know, I know,' he grumbles.

They gather a small hamper of items and bundle them together. They both put on large raincoats and Semyang finds an old umbrella near the front door.

'Are you ready?' she asks.

Tshering nods.

They walk out into the torrential rain together, the umbrella providing little protection. They follow the road carefully up the hill, slipping in the mud and looking at their feet to try and keep the rain out of their eyes.

The umbrella crumples and snaps in the wind but Semyang still holds it over them stoically.

'Just throw it away?' Tshering calls to her.

'What?' she asks, struggling to hear him over the noise of the rain.

He points to the umbrella and gestures tossing it away. 'Get rid of it,' he yells.

Semyang follows his suggestion and they continue their struggle up the hill together. They grasp each other's hands tightly, at times risking pulling each other over as they slip and slide but appreciating the human link in this devilish storm.

Heads down against the rain, neither of them sees the landslide as it careers down the hill towards them. The sound of the mass of mud shifting and plunging down the slope is masked by the cacophony created by the storm.

They are consumed by the mud and eventually suffocate, but even in their pain and suffering, they never let go of each other's hands.

SARAH

CHAPTER 5

Newton and I head back towards camp and as we approach, we hear muffled hammering. The Alphas, with the exception of Nat, are still sprawled around the fire pit.

The hammering appears to be coming from inside the Alpha hut. Newton and I head towards our own hut and duck inside. The interior is dark except for where light sneaks between some small cracks in the wood. Karim and Cassidy are sitting inside, and they nod to us as we enter.

Karim is showing Cass a monitor screen that he has salvaged from one of the chemo-pods, and he continues an explanation for her.

'...and by running wire from the pods to the huts, I should be able to power these and give us some lighting.'

I notice that Karim has more monitors in his lap.

'And the pods can provide the electricity?' Cass asks.

'Exactly. Each pod was installed with a small geothermal generator, so that the pods could monitor the environment and wake us when they needed too. I should be able to harness that power. The generators are only designed to give a small voltage, but that's all we need for a few LEDs.' He turns his attention to us as we sit beside them. 'You guys alright?'

'Fine, thanks,' I reply.

As we sit and chat Newton completes a few rudimentary tests with the new mushrooms that Ayesha found. I feel infinitely more comfortable with these three companions than when I'm with the Alphas. Even though we've only known the Epsilons for a few days and I've only known Karim for a few months, I can feel the strong bonds that we're forging.

'Does anyone have any plans for dinner?' I ask.

'Do you ever stop thinking about food, Sair?' Cassidy says in mock exasperation.

'Very occasionally,' I say with a small smile.

'I don't think so,' Karim says, answering my original question. 'Maybe check with Nat.'

I crawl out of our hut and head towards the Alpha's hut where the hammering continues. Ayesha sees where I'm headed and rushes to cut me off.

'Nat said to not let anyone in until she was finished,' she says, stopping me from going inside.

'Alright,' I say to Ayesha before calling to Nat. 'Do you wanna cook dinner, Nat?'

'Busy!' she calls, without pausing her hammering. I'm surprised she can see well enough in the dark hut, but I guess her eyesight has probably adapted to the dim light.

I scrounge through our dried food supplies and find a few packets of pea and ham soup. Based on the quality of our other freeze-dried meals, I'm not expecting them to have much flavour left.

I boil a large pot of water and add the powdered soup along with chunks of the large mushrooms we've foraged.

Ayesha has stayed beside me, and I point to the smaller mushrooms she found with Nat earlier on.

'What did you decide about these?' I ask her.

'Newton's tests came back clear so we're pretty confident they'll be fine. Still, only one person should eat them first. I'm happy to have them,' she says.

I pour a small portion of the soup into another pot, add some of the small mushrooms and place the pot on the edge of the fire.

Ayesha's soup gives off a wonderful smell and it draws Karim out from our hut.

'Something smells good, Sair,' he says with a grin.

'Not for you, unfortunately,' I reply. 'We're stuck with the flavourless fungi. Ayesha has offered to try the new mushrooms.'

'Didn't you say they're safe, Ayesha?' he asks, leaning over the pot and breathing deeply.

I swat him away. 'You know the rules. Now go and get the others, would you, it's almost ready.'

Newton and Cassidy emerge from their hut, but when Karim tries to tell Nat that dinner is ready, he gets the same one-word response. 'Busy.'

We were each provided with a basic metal cutlery set, a bowl and a cup in our original survival kits. I pour the soup into the bowls and hand it around the circle. Ayesha eats gleefully, directly from her pot.

'Delicious,' she says, teasing us.

I try the soup. The thin, green fluid is salty but flavourless, and the mushroom pieces are slimy and rubbery. I almost gag as I try and swallow a large piece.

'Better leave the cooking to Nat,' I say. 'Sorry, guys, this is our worst meal yet.'

'It's not that bad,' Newton says, chewing and chewing on a piece of mushroom of his own.

Jay has to cut the mushroom into much smaller pieces for Tomohiro to swallow whole, and I notice Cassidy drinking the soup but avoiding the fungus.

'The important thing is that you tried,' Declan says. I can't tell if he's having a dig at me, but for once, he seems genuine. 'I didn't see anyone else offering to cook, so thanks, Sarah.'

I'm a little thrown, but smile graciously and return to my soup, trying to get as much of it down as possible.

'Ahhh,' Ayesha says, draining her pot and smacking her lips.

'Rude.' Karim looks at her grumpily.

'I'm going to go for a run in the morning,' I say, trying to change the topic. 'Anyone want to join?'

'What's the point?' Declan asks, returning to his adversarial role. 'We're getting enough exercise as it is. You're just wasting energy.'

'Because I like running, and because I don't want to get out of the habit,' I respond tersely.

'I'll come,' Ayesha says. She smiles at me softly and I wonder if this is her way of apologising for voting against me earlier.

'Lovely. Anyone else?'

No one else is willing, so Ayesha and I agree to wake each other in the morning.

Nat emerges from the Alphas' hut and walks towards the fireplace.

'There's some soup left in the pot if you want it,' I tell her. 'But I'm sorry to say it's pretty horrid.'

'Maybe later,' she says with a small smile for me. 'Are you guys ready to see what I've been working on?' she asks.

Everyone murmurs agreement.

'Then, Ayesha, I need you to come with me.'

With a bewildered look back at the group Ayesha follows Nat to the hut. A few seconds later they re-emerge, Ayesha sitting in a small wooden vehicle and Nat pushing it along from behind.

The body of the vehicle is a flat A-frame with wooden planks attached to fashion a seat. Underneath, Nat has drilled four large holes into the supporting planks, through which she has passed two straight branches. To each of these branches she has attached two circular wheels, made from cross sections of a large tree trunk. The wheels at the back are large and thick, while the front wheels are smaller.

She pushes the vehicle up to Tomohiro and Ayesha hops out.

'Your chariot awaits, sir,' Nat says.

The wheelchair that Nat has created does look more like a chariot than a chair, but will hopefully function appropriately to give Tomohiro more independence.

The look on Tomohiro's face is pure joy mixed with gratitude. He may even be fighting back tears.

'Nat, this is…' he begins, but words fail him.

Declan and Karim help Tomohiro into the chair. Tomohiro immediately slumps to one side, and Declan catches him, preventing him from toppling off the chair.

Cassidy rushes off to her things and emerges with some long strips of cloth. She ties one around Tomohiro's waist and one around his chest, securing him to the chair.

Karim rests Tomohiro's limp left leg on the front of the chariot and Declan puts Tomohiro's left hand in his lap.

'Unlike normal wheelchairs, the wheels on this one are linked to each other,' Nat explains. 'That means it'll be harder to turn, but it should also mean that if you push with just your right hand, you'll go straight. Hopefully.'

Tomohiro pushes on the wheel and slowly but surely the chair starts to move. Once he has some momentum, he is able to keep going easily enough.

'I figured that to turn, you'll have to plant your good leg,' Nat continues.

Tomohiro does as she says. By anchoring with his leg and pushing with his arm, he pivots slowly.

'Sorry, it's a bit cumbersome and slow,' Nat says.

'It's amazing!' Tomohiro says with a huge grin. 'Come here, I need to tell you something.' He gestures Nat closer and when she is near, he places a huge kiss on her cheek. 'You're the best.'

Tomohiro spends a few more minutes practising in the chair. Though it's cumbersome, the large wheels handle the rough terrain well enough and he can manoeuvre mostly independently. Only occasionally does he get stuck and has to call for someone to help.

Of course, the other boys also want to have a turn, so after Tomohiro's arm gets tired, they each try the wheelchair out. With two functioning arms, they can get more speed and have more control than Tomohiro.

He sits to one side, watching them with a small amount of jealousy and sadness. Although Nat's present has lightened his spirit today, this surely isn't how he would've imagined his R.E.Generation experience to be. He puts on a brave face for all of us, but in this moment, I glimpse the grief inside him. I realise it will be up to all of us to not let that grief overwhelm him.

The next morning, instead of rolling over and snoozing when the sunlight breaks through the cracks in our hut, I get up as silently as possible. Grabbing my clothes pack, I head outside.

Karim, who was supposed to be on guard duty, is leaning against the outside wall of our hut and is sleeping soundly. I consider kicking him awake but decide to let him rest as the others should all be up soon anyway.

I rummage through my clothes, looking for appropriate attire. In the few days I have been awake, I've already worn most of my clothes, the near totality of my worldly possessions. I sniff the armpits of my cotton undershirts and choose the smelliest one. I have only one pair of shorts and, as the morning air is mild without any chill, I decide to leave these and stay in just my underwear. I pick up one of my two bras and glancing around to check that Karim is still asleep and no one else has risen, I take off my sleeping top and change in the open. The bra is supportive, designed for sportswomen, and as I do it up, I consider the moment in a year or so when the bras that I was allowed to bring with me have worn out, and I'll be unable to buy any more. I thought that the main challenge of survival would be getting set up, finding the essentials. But the challenges will be never ending. We're so used to fast fashion and fast technologies, but in this new world, when something breaks, if we can't make a new one; it'll be gone, and we'll have to learn to live without it.

I sneak into the Alpha hut and peer through the dimness, looking for Ayesha. Declan and Nat are snuggled up under the one blanket, Tomohiro and Jay are near the far end. Ayesha is closest to the entrance, so she's able to extricate herself easily after I wake her.

Ayesha changes while I visit the toilet. Like me, she opts to run in her underwear and a cotton top.

We set out on an unspecified route, heading roughly northwest and climbing uphill so that we get to run downhill on the way back.

'I wish I'd brought a sports bra,' Ayesha says as we settle into our rhythm. She is a few cup sizes larger than me.

'You didn't bring one?' I ask, surprised.

'The packing list said two bras, I didn't think about the different types.'

'And now you're stuck forever without a sports bra. I'm sorry.'

She shrugs and holds her arms across her chest to support herself.

We work up a sweat as we climb steadily, but the wind helping to push us up the hill is bitingly cold.

'Feels like a cool change,' Ayesha comments.

'Gee, it's fresh, isn't it?' I respond.

After five or ten minutes, we turn to our left, cutting obliquely across the slope. Now that we're no longer climbing, the wind cuts deeper, the exercise only just managing to counteract the decreasing temperature.

A flicker of motion up ahead momentarily distracts me from the weather.

'Did you see that?' Ayesha asks.

'Yeah. What did you see?' I ask.

'It moved too quickly. I couldn't make it out. It was just a blur, really. Scales, maybe.'

'Yeah, me too,' I reply. By the time I'd turned my attention to the movement, it was already gone. But a dark flash definitely disappeared behind a large tree.

We approach the tree cautiously, but there's nothing behind it now. Cassidy's words echo in my brain: *nobody ever thinks to look up.* I crane my neck but can't see anything out of place in the branches either.

Ayesha's teeth start chattering and I realise I've also cooled down further since we stopped. While Ayesha has bigger breasts than me, she's also skinnier. I would curse her good luck, if I didn't know that she was a bit of a fitness freak beforehand, and that she's worked hard for her figure.

We head off again but soon it starts to rain, and we decide it's prudent to return to camp expeditiously. The rain is freezing and the wind blows it directly into our faces. It soon turns to hail that stings as it cuts at our bare arms, faces and legs. The storm builds in severity as we jog, then stumble back towards camp. The ground underfoot rapidly becomes sticky mud and I need to help Ayesha back to her feet after she trips on a stray root. Her teeth chatter uncontrollably now and, despite the mud on her hands and knees, I hold her close to try and share my body heat with her. Huddled together, we struggle towards camp.

The foul weather comes in aggressive sheets, blown in by the ever-increasing wind. The pine trees offer no protection, bent double by the gale and the rain, driven almost horizontally.

With my head down to protect my face from the stinging ice,

my sense of direction is a bit off, however, I manage to direct us to the camp after a few adjustments. Ayesha and I are both shivering uncontrollably as the others wave us towards shelter.

I take us to my hut. Newton rushes past us as we arrive, braving the storm for some unknown reason.

Cassidy makes us strip-off and lie down. She covers us with two blankets each, the only bedding we have inside this hut.

I blackout from the cold and the following hours come in broken moments as I pass in and out of sleep.

I feel Cassidy slide under the blankets to share her warmth with me. I hear steel striking flint and a man cursing as the wet kindling refuses to catch.

I wake, shivering uncontrollably and aching all over. I cough, causing more pain to course through my aching body. The air is full of smoke, and firefight dances on the roof of the hut. Somehow, I'm wearing clothes again, so Cassidy must have dressed me.

I doze again, waking to the sound of hail thundering on the slanted walls, and small droplets of water dripping near my head. The room is warmer now, the air thick with smoke.

I slip into a deeper sleep and lose all sense of time as exhaustion grips my body. My life continues in fragments and I can't tell if hours or days pass.

Cassidy feeds me flavourless soups and warm water to help maintain my internal temperature. The others must be keeping constant watch over me, because each time I wake, someone is always sitting beside me with a friendly smile.

At some point, the rain and hail stops, but anything outside the aches of my own body and my dealings with my carers barely registers with me.

I feel my strength and clarity returning and I'm able to talk to Cassidy. She looks shattered. Exhausted both emotionally and physically.

'It's alright, Cass,' I manage to say. 'I'm gonna be fine. You can stop stressing.'

The anguish in her eyes doesn't lessen as she looks me over. 'It's not you I'm stressed about. It's Ayesha. I'm not sure she's going to make it. She hasn't woken up yet, so we can't give her hot fluids.'

'Shit,' I mumble. In my own bubble, I'd totally forgotten that she'd even been with me in the storm. It was my idea to go running. If she dies, that's on me. Just like Tomohiro's paralysis.

Soon after, the others carry Ayesha from the hut. They don't talk to me about her and I assume the worst. Likely, they're shielding me from the truth until I'm stronger.

My body temperature appears to have corrected itself, but I'm still exhausted, so I stay in bed a further twenty four hours. As I languish in bed, dark thoughts creep out from the recesses of my brain and I torture myself with dreams and nightmares.

Even though I assure them I'm fine, the others continue to keep me company. Cassidy, Newton and Karim ensure there is always someone sitting or lying beside me. Declan even sits with me on one occasion. We talk about idle things, where we studied, plans for the future. That sort of thing. Now that he's in charge, he seems to have relaxed, and his good qualities are coming to the fore, replacing his character flaws.

Still, no one mentions Ayesha and I don't have the heart to ask. I try to pluck up the courage time and again, but never manage to get the words out.

Nat is clearly constructing something else outside as the sounds of sawing and hammering once again fill the forest. These noises are interspersed with curses and shouting.

You imbecile! and *stupid, bloody cripple*, appear to be her new favourite phrases.

When I ask Karim about it, he explains that Tomohiro is trying very hard to be useful. Now that he's able to move around in his wheelchair, he's offering help, however he is clumsier with only one hand and reduced core strength and Nat doesn't appreciate his efforts.

I spend that afternoon lying in bed staring at the wooden structure around me. My eyes have adjusted to the constant dim light, so I can clearly see most of the markings. The knots in the wood make fabulous patterns and my brain finds many faces and shapes amongst them.

I point them out to Newton, and he lies down beside me to find his own. The challenge becomes trying to find a more interesting picture than the previous ones.

'There's one,' he says pointing at a collection of whirls. 'It looks a bit like a bear.'

'Ah yes, I see it.'

Two knots represent the ears and another the bear's fluffy nose. It also has two small dots for eyes, although the left eye is a bit out of place, giving it a lopsided appearance.

The next day, I'm able to get up and go outside. When I first slip between the wooden panels that make our entranceway, I'm blinded by the light outside, however my eyes quickly readjust to the sunlight.

Nat has started a small, traditional hut that Declan proudly shows me. They decided that as leader, he should be afforded his own residence. While this goes against what they emphasised about cooperation in training, it seems true to form for Declan, and I'm not going to fight him on it. The design is sturdier, with the four large corner posts dug into the ground. She has started on the walls which are made by overlapping planks, like a clinker-built boat. The hut will have a slanted roof, a wooden door and a small window cut-out, to let light in. Inside, Nat is planning on laying wooden planks to make a floor. It's only about fifteen metres square, but it should be a cosy building.

I use a small amount of water to refresh my pits and bits, but it seems futile. To this point we haven't been able to afford water to wash any of our clothes, so collectively, we smell like the worst boys' changeroom that you could imagine.

Over the next couple of days, we finish Declan's house and set about making two more huts on a larger scale. We discuss making multiple small huts, but most of us decide that we're happier living together for the moment. In time when we start having families, we will need to build more homes, but for now the camaraderie and company are more important.

Nat, as always, takes control of construction. Tomohiro plays a key role in choosing the trees to be felled, the team needing to venture further from camp in order to find appropriate timber. Much to Tomohiro's dismay, he is unable to help cut the trees down, but he spends large portions of the day stripping the bark and branches from the fallen specimens. The process creates lots of waste timber, which is collected into a large pile and feeds our ever-hungry fire.

I'm unable to be much assistance on the first day that I'm up. Every time I stand, I feel lightheaded. It's amazing how, after just a few days inactivity, I feel noticeably weaker, possibly due to muscle wastage or just lingering fatigue. Declan also appears to be shirking work, disappearing into the Alpha hut at regular intervals.

Twice a day, Newton climbs the tree that we originally climbed together. The others have taken to calling it the *weather tree*, and he provides a report after each climb. He hasn't seen much more of interest, but they're hoping he'll see any other large storms in time to warn us.

From the second day, I focus on rebuilding my strength and can help ferry timber, dig foundation holes and hammer boards in small bursts. I take regular rest breaks, sitting on one of the many exposed stumps that litter the clearing, marking the locations of felled trees.

Taking a breather, I watch Tomohiro sitting in his chair and stripping a large log that has been hauled up beside him. Luckily, he retained function in his dominant side, so he can still use the small hand axe. He cuts down between the log and his chair, taking a couple of blows to remove each branch. Nat approaches him from behind to inspect his handiwork. I watch as Tomohiro swings his axe and, meeting less resistance than expected, cleaves straight through a small branch and into the toe of Nat's boot.

She releases an almighty scream that echoes around the forest. If there were any birds around, I imagine they would've taken flight in fright.

'You fucking, lumbering, stupid, bloody oaf!' Nat rages.

'What the hell are you doing sneaking up behind me,?' Tomohiro retaliates.

He withdraws the axe and Nat slumps onto the log, lifting her foot into her lap.

I'm halfway towards them before I even realise that I've stood up and I fight back a dizzy spell as I approach. I can hear others approaching from the distance, but I'm easily the closest.

Nat takes her boot off and her sock is filled with blood. We carefully peel that off, too, to reveal the damage underneath.

The axe has sliced the webbing between her first and second toes and cleaved about a centimetre into the muscle.

'You're quite lucky,' I say. 'There's no bone damage and not much structural damage either.'

'No thanks to him,' Nat says, gesturing to Tomohiro aggressively.

'Leave him be,' I say. 'How could he know you'd just stepped behind him? You need to take responsibility for your part in this.'

'Maybe, but it bloody well hurts, alright!'

Most of the others have gathered and Declan and Cassidy inspect the wound before leaving me to clean it and stitch it closed. After I've closed the wound, I bandage the two toes together.

I explain the recovery process to Nat. 'Normally, we'd give you antibiotics to prevent infection, but with limited supplies, I think we should wait and see how it goes. Keep it elevated today to minimise swelling and no walking on it for a week, okay?'

'No way. I need to build these huts,' she replies.

'You've shown us all the basics, and you can still project manage, but seriously, you have to stay off it so that it can heal.'

She grumbles but accepts my advice. 'Can someone build me a wheelchair?' she asks.

'You'll have to share Tomohiro's,' I say with a smile.

Tomohiro relinquishes the chair and I help him to the fireplace where he slumps against a stump. The effort takes a lot out of me and I wish we had three wheelchairs.

'You alright?' I ask.

'Yeah, fine,' he says with a forced smile.

'No, seriously, Tomohiro. How are you?'

'I was feeling a lot better after Nat gave me that chair, but ever since, I've been trying to make myself useful and she just keeps yelling at me.'

'I've noticed,' I say.

'And now this,' he says, referencing the accident. Talking remains effortful for him with his half-paralysed face.

'You know that wasn't your fault, yeah?'

He nods.

'Good. Ignore Nat. She's being unreasonable. She's in pain but she'll calm down. And the rest of us respect your efforts.'

'It just sucks, the other Alphas are so close. It feels like Nat and Declan are pushing me out.'

'You're a great guy, Tomohiro. If they can't see that, that's their own stupidity. But if you keep proving yourself, keep improving, they'll come around.'

He smiles at me. 'Thanks, Sair. You're a gem.'

He looks genuinely calmer. I give myself a metaphorical pat on the back.

The others continue to build the large huts, with Nat bellowing instructions and regular abuse.

I take a walk into the forest to stretch my legs and build more strength. Since the storm, the weather has been warm and humid, and I notice that new mushrooms are emerging from the forest floor. It appears that there are multiple types of fungus appearing, but I can't be sure. I wish that Ayesha was here. She'd be able to identify the differences.

Multiple times I've got close to asking what happened to her, but the words stick in my throat and my resolve fails. I expected to feel a collective sorrow from the group, but everyone simply seems to be getting on with surviving.

While I'm bent over inspecting the new fungi, I suddenly hear a loud scurrying sound from behind me, but by the time I've turned around, there's nothing there. I scour the trees, but again, whatever made the sound is either gone or hiding.

I head back to the camp, making sure I don't tread on too many of the baby mushrooms, which is surprisingly difficult in some areas where the entire forest floor seems covered in fresh growth.

I take an early night, feeling drained from the days exertions and wake feeling almost myself again.

That morning, while I'm hunched over my bowl of gruel, Newton looks over my shoulder and smiles.

'Look who decided to get up?' he says.

Turning around, I see a drained, but very much alive Ayesha, leaning on Declan as he escorts her to the fire.

'Nice of you to join us,' Tomohiro says sweetly.

'She's alive?' I ask in indignation.

Everyone turns to look at me. Eight faces showing varying degrees of confusion and amusement.

'Of course she is,' Nat says.

'But Cassidy said she didn't think she'd make it, and then you all carried her away,' I say.

Declan shrugs his shoulders. 'She wasn't getting better under Cassidy's care, so I had her moved to my hut so I could keep an eye on her.'

'But no one told me she was alive,' I say indignantly.

'I thought Cass would've told you,' Newton says.

'I didn't realise I had to,' Cassidy says.

'No one told you that she'd died, either,' Karim says. 'I think we might've mentioned that.'

I rush forward and give Ayesha a big hug. 'Well, you're alive,' I say. 'That's the main thing. I'm so, so sorry.'

'What for?' she asks.

'Taking you running in that storm.'

'We didn't go running into the storm, Sarah. It came out of nowhere. It wasn't your fault.'

I give her another hug and take her to the fire. 'If you feel as bad as I did two days ago, just take it easy.'

'I ache all over,' she replies with a faint smile.

'I know what you mean. But it'll get better.'

FRANCISCA

Francisca is woken from her reverie by a loud knock on the front door. When she is cooking and cleaning, she goes within herself to block out the racket created by her four children. These days, people would literally kill to have a family like hers, but even so, Aline, Maria, Lucas, and Antonio sure know how to push her buttons.

'Children, I need you to be quiet for a few minutes,' she calls. 'There's someone at the door.'

She crosses their humble hut and pulls the door open. She is surprised to see a huge, white man in a fancy suit standing in front of her. People like him don't typically frequent their small town on the Brazilian Plateau.

'Sorry, I think you've got the wrong house,' she says, going to close the door.

'He can't understand you,' a lady says, poking her head around the man. She is about half his size and looks like she could be a local. 'He only speaks English,' she continues. 'But I'm here to translate for you. May we come inside?'

Francisca lets them in nervously. The man wears sunglasses, which he removes as he enters the house. He has to duck to fit under the door frame.

'Please, take a seat,' she says gesturing to their old couch. 'Would you like something to drink?'

The lady translates and the giant man asks for a cup of tea.

'You want tea, even when it's this hot?' she asks.

The man nods.

Francisca makes a pot and then sits down opposite the odd couple.

The man starts to talk, and the lady translates for him.

'Sorry to interrupt your day. My name is Milo, and this is my interpreter, Camila. I am from NAMA.'

'What's NAMA?' Francisca asks.

'We're the organisation who took over from NASA once they went broke. But we focus entirely on Mars, rather than the whole of space.'

'And why are you in my home?' Francisca asks, still confused.

'I am here to invite you to join our mission.'

'Why me? I don't understand.'

'Not just you,' Camila translates. 'All of you. Your husband and children too.'

'Okay then, why us?'

'We have completed stage one of our Martian mission, which was to build an inhabitable enclosure on the Martian surface. A team of robots have created a new home up there, away from the troubles of this world, and we're looking for people to inhabit it.'

'Why us though? Why have you come all the way here?'

'You have a proven track record. Your fertility is guaranteed.' Milo motions to her children. 'In America, we have only had one-thousand healthy births in the last three years, and only two of those are not the first-born child of their family.'

'I knew *The Blight* was bad, but I didn't realise it was that bad,' Francisca muses.

'It's the same all over the developed world.'

'So, you want us to live on Mars?' Francisca asks.

Camila translates and Milo nods.

'May I ask, if you've managed to create somewhere for humans to live safely on Mars, why can't you build an environment like that here on Earth?'

'We're trying that too, but the Earth is very volatile at the present time. There's a lot of belief that we should play the odds by trying all options.'

'So, you think it could be safer to live on Mars?' she asks incredulously.

'The world we live in, hey?' Camila says, without bothering to translate for Milo.

'May I have some time to discuss this with my husband and family?' Francisca asks.

'Of course, take all the time that you need.'

Two weeks later, Francisca, her husband Jose and their children fly to America. They have to complete some basic training and are introduced to the other families that have also agreed to go on the mission. They come from all over the world and speak all different languages. They have lessons to learn a new common language called Martian. It's an incomplete and basic language, but has all the vocabulary required for the mission.

Once everyone is ready, they must wait for a calm day to launch the shuttle. They have a few false alarms, and on one occasion, they are all strapped in and waiting for take-off, before the launch is cancelled due to strong winds.

The progressively worsening weather has made it harder and harder to launch anything into space. One of the other volunteers tells Francisca that NAMA lost almost fifty percent of their rockets that took the terraforming robots to Mars.

'Then I'm very happy for them to wait for an extra calm day,' Francisca says in reply.

Over the next month, she can see the NAMA workers getting increasingly agitated.

'Why is everyone so hot and bothered?' she asks Milo, through Camila, when they come to check in on Francisca and her family.

'It's coming to the end of the dry season and it's much harder to launch in the wet,' he replies. 'They had expected to have had a good launch day by now, but the Earth isn't playing fairly.'

The following day, they are called to the launch site and Francisca gathers her children and things together. As they approach, she can see that it is windier outside than the day that they cancelled the 'almost' launch. *Oh well, maybe tomorrow*, she thinks.

When they arrive however, the NAMA staff fuss over them and they are herded onto the space shuttle with little ceremony. They are strapped in securely and the staff complete final checks.

'But the wind?' Francisca asks one of them.

'It'll be fine,' the lady technician replies. 'We've made some modifications so that we can launch in less favourable weather.'

'Have the modifications been tested?' she asks.

'Everything will be fine,' the technician says, before closing the door and locking them inside.

'Are we going to space, Mummy?' Antonio asks.

'We are, my darling. The greatest adventure. Are you excited?'

He grins and nods vigorously.

She smiles back at him but feels uneasy inside.

The shuttle vibrates violently as the launch procedure is initiated and they start to rise into the air. The G-force hits Francisca even harder than she anticipated. Suddenly, she feels unwell as she is battered with unexpected rotational forces. It's too much for her to handle and she blacks out completely.

The NAMA staff on the ground watch in terror as the shuttle spins violently. Barely off the launchpad, it seems to cartwheel in the air as it is buffeted by the wind, before abruptly bursting into flames.

Camila pats Milo's back as tears roll down his cheeks. They watch together as the volunteers and wreckage rain down onto the hungry earth.

SARAH

CHAPTER 6

Over the next few days, we complete the first large hut. Big enough to comfortably fit eight or nine people for sleeping. Under Nat's direction we add an internal wall, sectioning off a small changing area. I assess my hands which have a number of blisters. They're not used to hard work, and I've had to push through the pain of them popping to keep helping the others. I can only hope that they solidify into calluses sooner rather than later.

Looking at the rough timber floor of the hut, I wonder if it's going to be an improvement. The forest floor of dirt covered in pine needles is comfortable enough and I've slept well considering everything that we've had to deal with. The timber floors make the hut feel more homely, but don't look comfortable. Unfortunately, wishing for a mattress won't make one miraculously appear.

I check and clean Nat's wounded foot daily. She has managed to avoid getting an infection, something we're all thankful for. The wound looks healthy and heals well. Despite my protests, she starts walking again five days after she was injured. She hobbles around for a few days, but she's largely fine.

I collect some samples of the new mushrooms and take these to Ayesha and Newton. Based on Ayesha's knowledge and simple titrations completed by Newton, they sort them into piles that they think are dangerous and hopefully safe. There are five new types of fungus and they identify three as edible. A bright orange one, a squat, pale brown mushroom and a small type that looks like a conical Asian hat on a thin, dark-brown stalk. This cute mushroom grows in large groups on fallen sticks and logs.

A black, leathery mushroom and an inconspicuous white button-like mushroom are rejected.

When it comes to testing the mushrooms, Tomohiro insists on trying them all.

'I'm the most expendable,' he says firmly.

We try to argue with him, but he is adamant and so he alone tests one each day. We monitor his condition closely, but he shows no negative symptoms.

'They're all delicious,' he tells us after trying the final mushroom. 'This one is nutty, the small ones are a bit spicy and the orange guy was really mushroom-y.'

We're all thrilled to have something with more flavour entering our diet.

Someone suggests creating a map of the mushroom's locations, which turns into an idea of mapping the entire route to the spring, and so we plan a large expedition to our water source. Although we each arrived with fifty litres, our stocks are starting to run dangerously low.

We decide to split into two groups, an advance team to carry our large water canteens to the spring and a rear guard fanned out to search for mushrooms and relay information to Ayesha to record.

'Karim and Sarah have been to the spring previously,' Declan says, taking charge. 'So, they should be split across groups to act as guides. The next strongest pair would be me and Nat, so we'll go to the spring with Sarah and the rest can stay with Karim and Ayesha.'

'Who says that you two are the strongest?' Newton asks, irritated. 'Jay and I could both give you a run for your money.'

'I say, and as leader, that's that,' Declan responds.

'That's not the power that we agreed the leader would have,' I say carefully.

Declan returns my gaze with a stony stare.

'Are you okay going with Declan and Nat?' Newton asks.

I'm touched that he's worried about my safety, identifying a potential threat that I hadn't even noticed, but seeing as Declan has been a lot better recently, I don't feel at risk.

'I'll be fine, I can look after myself,' I say with a wink.

'Couldn't Karim go with Nat and Declan?' Newton asks, pushing the point.

'Let it go, Newton,' Cassidy says, putting a hand on his shoulder.

'Listen to your girls, mate,' Declan says patronisingly. 'And don't forget what I told you before the election.'

He says it with a smile, but the threat doesn't go unnoticed.

A week and a half after the storm, Ayesha and I are back to full health and Nat insists her foot is up to the hike. Newton climbs the weather tree and reports that the skies are clear.

Ayesha, Newton, Karim, Cassidy and Jay set out towards the spring, walking in a wall with a five to ten metre gap between each person. They call to each other, reporting what they find. Soon, their calls are lost in the forest.

Tomohiro has agreed to stay behind, nominally guarding the camp. Declan, Nat and I take his wheelchair and tie our large canteens onto it. We pack smaller canteens into bags and sling these onto our backs. Once we're ready we set out and soon overtake the others who are progressing slowly, mapping the location of the different edible fungi. As we pass them, Ayesha shows me a new type that they've found that will need testing.

The wheelchair is a bit awkward to wheel through the forest, but we fall into a rhythm and chat idly as we walk. Nat and Declan are pleasant, and we progress well.

When we reach the small cliffs, we manhandle the wheelchair up without much difficulty, or else find a short detour around them. A few kilometres from the spring, I notice a change in the others. They stop chatting and nervous looks pass between them as unspoken messages are sent and received.

'What are your thoughts on Tomohiro?' Nat asks me, breaking the silence.

'He's a great guy,' I reply. 'I'm not sure that I would've coped if I'd been in his situation.'

'You tell her,' Nat says to Declan.

'There's actually a reason we wanted to be alone with you, Sarah,' Declan begins.

I feel myself tense up automatically. Maybe I should've pushed to have Newton with me after all.

Declan continues. 'As unofficial leader of the Etas and Epsilons, we need you onboard.'

'Onboard with what?' I ask.

'Tomohiro is a liability. Look what he did to Nat. And he's a drain on our resources.'

'So? He's contributing, just like everyone else,' I reply.

'Not enough,' Declan says.

'We think we should get rid of him,' Nat adds.

'Sorry, what?' I say dumbfounded.

'We can't afford dead weight in our society. It makes sense to euthanise him,' Declan says coolly.

'You're seriously talking about killing Tomohiro?' I ask incredulously.

They nod.

'You're a sensible person,' Declan says, trying to win me over. 'If you think about it, it's the logical decision, but we need everyone to agree. You have some power over the others, so if you agree, they'll be on board.'

I shake my head at them. 'I'm sorry, but I'm not who you think I am. There is no way I can condone something like that. Tomohiro has so much to give. He's testing all our food for us, he can sew, he can cook. He can have kids. He risked everything to volunteer for this project, and unlike many of the others, he made it. He risked everything again with that grub, and lost the use of half of his muscles for it. That doesn't mean he deserves to die. That means he deserves to live.'

They look at me, disappointed.

'I told you she'd be like this, Declan,' Nat says.

'Well, it was worth a shot,' he says, turning away and continuing up the hill.

'You said that we all need to agree,' I say, rushing to catch up with him. 'You're not gonna do anything, are you?'

'Of course not,' he says, but he doesn't look at me.

We walk in silence the rest of the way, taking it in turns to push the wheelchair.

Nat and Declan are both excited to see the spring for the first time. They grin like children and rush towards it once we get close. They drink excitedly and take the opportunity to wash themselves and their clothes. After drying off for a few minutes, Declan tells us that he's going to go check on the others to see how far they've progressed.

Nat and I fill all the canteens, strapping them securely onto the wheelchair.

We sprawl on the ground to wait for the others, not talking to each other. I don't dislike Nat as much as I dislike Declan, but we just don't have anything in common and there's nothing for us to talk about.

I wonder what it is that she sees in Declan. Of all the Alphas, she is his most loyal lieutenant. Unlike Ayesha and Tomohiro, who I'm sure follow Declan because of fear, Nat seems to follow him because she genuinely likes him. Maybe she's cleverer than I give her credit for. Maybe she has simply hitched her wagon to power, to give herself the best chance at a comfortable life.

Looking across at her, I wonder what she's thinking about. Is she thinking about me, like I'm thinking about her? She's watching the water as it trickles past. Maybe she's just thinking about that, or maybe she's thinking about who she left behind to volunteer. She glances up and catches me staring. I look away awkwardly and try to find something to do with my hands. I glance back at her. She has a serene look on her face, but nothing to hint at what is going on inside her mind. I guess I'll never know.

With time to wait, I get lost in my own thoughts. Normally, I try not to think about 'before', my family and what I had to give up to join this project. It's easier to stubbornly think about the future and the challenges we continually face. It's bizarre to think that my parents and my sister, Alice, ceased to exist more than fifty-thousand years ago. Did Mum and Dad get to go on their dream holiday to Hamilton Island for their fiftieth anniversary? Did Alice ever find someone to marry who lived up to her exacting standards? More questions I'll never know the answer to.

After waiting fifteen minutes, Nat turns to me. 'Do we need to wait for the others?' she asks.

'Well, we agreed to meet here,' I say. 'Plus, everyone else will be excited to see the spring and will want a wash.'

She begrudgingly agrees and we wait another ten or twenty minutes before we hear them approaching.

I walk towards them and Cassidy gives me a big smile. 'We made it,' she says.

'Where's Declan?' I ask, looking around the group and not seeing him.

'What do you mean?' Karim asks. 'Isn't he with you?'

'No, he said he was going to check on you guys.'

'Well, we didn't see him,' Newton says.

A cold shiver runs down my spine as I remember the conversation that we had on the way to the spring. 'Tomohiro,' I say with trepidation.

Without another word I break into a run, heading back for camp. I hear people following me and soon Cassidy and Newton are beside me.

We run in silence, racing down the hill and sliding down the cliffs. I have to pace myself, holding back from sprinting the first kilometre, knowing that I still have six more to go.

Declan, however, had a half-hour head start. He's probably already back at camp. What will sprinting back achieve anyway?

As I run, I picture all the ways Declan might euthanise Tomohiro. What will we find when we get back?

After about five k's, despite my best efforts to restrain myself, my heart burns and my legs ache. Cassidy lopes along beside me, seeming to manage the run easily, her longer legs eating up the metres and her fitness levels higher than mine after my sickness. Newton's face is flushed red, but he, too, is fitter than me and coping fine.

As we near the camp, I brace myself for what I expect to find. We race down a hill towards the Epsilon pods, still a few hundred metres from camp.

Cassidy and Newton's pods are closed, protecting the inside from the weather, but the third pod, that should still be buried in the hill, is open.

We slow to a walk as we approach, trying to catch our breath.

'The fuck?' Newton asks.

The pod has been excavated from the hillside, and once we're close enough, we see a skeleton inside. Closer inspection of the chemo-pod shows a large dent that must've broken the seal.

'I'm so sorry guys,' I say.

'It has to have been Declan, right?' Cassidy asks.

'He threatened them when we said we were gonna vote for Sarah. I guess this is because I questioned him about the hike,' Newton says.

'Or because I didn't agree with him about killing Tomohiro,' I say.

They both turn to me, dumbstruck.

'Nat and Declan wanted to euthanise Tomohiro because they think he's a burden,' I explain.

'That's revolting,' Cassidy says.

'Well, it looks like nature did the dirty work for him here. What would he have done if she was still alive?' Newton asks.

None of us answer, but I'm pretty sure I'm getting to know Declan well enough that the answer is clear. This wasn't just a threat. If the Epsilon volunteer had been alive when he opened the pod, I don't think she would've been alive soon after.

We continue on to the camp and are relieved to find Tomohiro resting where we'd left him. He has chopped up a large pot of mushrooms.

'Where's Declan?' I ask him.

'Haven't seen him,' he replies, looking up as we approach. 'Isn't he with you?'

'No,' Cassidy says simply.

We sit down to wait for the others. Cassidy takes the pot of chopped mushrooms and puts it on the fire. We have no butter or oil, so she has to stir them constantly to stop them from sticking and burning.

Cooking with the new mushrooms produces much more delicious smells. Soon peppery, nutty aromas fill the air. Cassidy adds a large amount of water and a few packets of a dried soup mix.

Jay, Nat, and Ayesha arrive back to camp. They enquire about my mad dash back to camp, but I answer vaguely. A few minutes later, Karim arrives, pushing the wheelchair laden with canteens of fresh water.

None of the others have seen Declan.

'I hope he's not lost,' Ayesha says innocently.

'Trust me, he's not,' I say curtly.

'But he could be,' she replies.

The meal is soon ready and we all tuck in, thanking Tomohiro and Cassidy for their efforts.

The soup is delicious. Probably the best meal we've had since we woke up.

According to Ayesha, the mushroom season may last a couple more weeks or maybe months, but at some point, we will need to find something else to eat.

As we're finishing our meal, Declan stumbles in from the forest, looking flustered and tired.

'Thank God I found you,' he says. 'I, I, I lost my way, and I thought I wasn't going to be able to find my way back to camp.'

I think his performance is a little over the top, but his acting is convincing and many of the others seem to be lapping it up.

'See, I told you that he might be lost,' Ayesha says petulantly.

'Why'd you do it?' I ask Declan directly, ignoring his charade.

'What, get lost? I didn't do it deliberately.'

'You know what I mean,' I say.

'I haven't the foggiest idea what you're talking about,' he replies.

'The Epsilon,' Cassidy says starring daggers at him.

Newton has his arms folded in front of him and a stony look on his face. Everyone else is looking between Declan and the three of us, bewildered.

'Pray do explain,' Declan says. His holier-than-thou attitude tells me all I need to know about his guilt. The others don't seem to have noticed the game he is playing.

I sigh. 'If you need us to spell it out, then follow us.'

Newton leads everyone to the location of the Epsilon pods. Having unloaded the water, Karim pushes Tomohiro in his chair to help him keep up.

I gesture to the open pod where the skeleton of the Epsilon volunteer lies.

'Why did you dig them up?' Declan asks. 'And what's this got to do with me?'

'*We* didn't dig her up!' Newton says.

'It was like this when we got back. You dug her up, didn't you?' Cassidy asks accusingly.

'Of course not. I was lost in the forest. How could I have excavated this?'

'What were you gonna do if she was alive?' I ask, ignoring his denial.

'He's told you he was lost. Leave him alone,' Ayesha says.

'Yeah, who's to say that you guys didn't dig her up?' Jay asks. 'Then after you found her dead, you've decided to blame Declan.'

'Why would we do that?' Newton asks.

'To even up the numbers again,' Tomohiro suggests with a shrug.

'We didn't dig her up,' Cassidy says.

'Well, if you didn't do it and Declan didn't do it, then who did?' Nat asks.

'Declan!' Cassidy and I say together.

'You've got no proof it was me,' Declan says.

'It could've been Tomohiro,' Karim says. 'You were alone and nearby, mate.'

'And without my wheelchair you reckon I could've got here and excavated a pod?' Tomohiro rebuts.

'Hardly,' Karim says. 'Sorry mate, it crossed my mind so I thought it was worth saying.'

'Could've been an animal?' Nat suggests.

'Sure, maybe an animal might've been able to dig the earth out, but no animal could open the pod,' Newton says.

'No animal that we know of,' Jay says. 'But hey, we don't know what's out here.'

'True,' Karim says.

'Well, the only people that we know that were here with the pod and had the time to dig it up are Sarah, Newton and Cassidy,' Nat says.

While everyone is turned towards the three of us, Declan smiles smugly.

'It bloody well wasn't us,' Newton says.

'And Declan says it wasn't him,' Ayesha says.

'I guess we'll all have to make our own conclusions then,' Declan says.

The group disperses and the Alphas head back to camp, leaving the three of us with Karim.

'You believe us, don't you?' I ask him.

'Of course I do. But Declan's clever, and like they said, there's no proof.'

'He is clever,' Newton says. 'I thought this was just payback for standing up to him and a warning to toe the line. Deliberate or not, by letting us blame him he's also managed to gather the Alphas closer around him and further ostracise us.'

'I'm not sure how much more of this I can take,' I say.

The others nod and we amble back to camp in a sombre silence.

BRAD

Brad's birth was greeted with huge celebrations. The first child born inside the bunker. A great hope for the future. The mood was much more muted the following day when his mother passed away from birth complications.

His father and the other adults did a wonderful job of raising him. There was always someone willing to play with him, willing to teach him.

After seven years had passed and no other babies had been born, the reality started to set in for the adults. Brad wasn't the first child born into the bunker. He is the only child.

The bunker itself is huge, with enough room to house twenty families and farming areas to grow food under UV light. Brad loves exploring the bunker, but by the time he is ten, he knows every nook and cranny intimately.

Brad doesn't know any different and doesn't know what he's missing, but the adults despair to see him playing on his own and missing out on exploring the great outdoors.

On his thirteenth birthday, everyone cheers as he blows air across the top of his cake. A strange tradition that Brad doesn't quite understand.

'One day, when you're older, you can leave this place,' his Dad says.

'But why would I want to do that?' Brad asks. 'I've got everything I want right here.'

The adults share their knowledge with him, and Brad becomes one of the greatest tradesmen of all time. He can farm, weave, sew, carve, and cook.

When Brad is thirty, Aunt Elise dies and it's the first time that he has experienced loss. Sure, they'd had animals die before, but that was always welcomed due to the meat they provided. He struggles to wrap his brain around the sudden disappearance of his Aunt. She was with him, and now she is gone, and they say that she will never come back.

Soon, Brad becomes an expert in grave digging as well.

He is forty-three when his father dies. He cries and cries, but nothing he can do will bring the great man back.

He keeps working, making sure that there is enough food for him and the few old codgers that are left.

The day before Brad's fifty-fifth birthday, Uncle Pat succumbs to a terrible cough that he's had for months. Uncle Pat is the last of the adults to die and so Brad is left on his own. Master of the bunker and his own destiny. He continues to farm the land. Chickens don't suffer from *The Blight*, so he keeps breeding them and they provide all the company he needs.

On his seventieth birthday, he thinks about the words his father said when he was only thirteen. *One day you can leave this place.* He shakes his head, remembering his answer and smiling. *I've got everything I need right here.*

Sure, the adults told him about the sun, and he is curious to see what it looks like, but the lights in the bunker seem to do a mighty fine job. He also wonders if there are any other humans alive out there, but having never met a new person, he doesn't even know what he would do if he did meet one. He was always on hugging terms with everyone in the bunker, so if someone came into the bunker now, he would hug them. That's just the done thing, right?

Soon after Brad turns seventy six, he feels a flutter in his chest. His shoulder aches and he starts to feel lightheaded. He sits down to breathe through the pain, but it only intensifies. He smiles before his heart gives out completely. Happy and content with the life he lived. Ignorant of the life he missed out on living.

SARAH

CHAPTER 7

The following morning, I take another run to help clear my head and sort through my feelings. I dare not ask Ayesha to join me and the others in my hut decline the invitation as well. Newton does, however, insist on climbing the weather tree, to ensure there is nothing on its way before allowing me to leave camp on my own.

I'm feeling much better after my illness, but I make myself ease into the run, taking small steps and plodding across the needle-strewn forest floor. My thighs ache after yesterday's run back to camp, so it's good to stretch my muscles.

I've always been a strong believer in democracy, but the strangle hold that Declan has manufactured in our small society has me questioning whether it will invariably lead to something else. I guess the corruption and intimidation is more obvious in such an insular community. It does make me see with new eyes the similar tactics used by leaders and political parties for generations.

My problem is trying to work out what to do about it. The original thirty two volunteers in our unit that were supposed to wake together and forge a sustainable community, have been reduced to only nine. If Newton, Karim, Cassidy and I were to separate from Declan and the others then we'd have absolutely no future.

Therefore, we need to stay with the Alphas for now, but if that is going to be sustainable in the long term, Declan needs to be dealt with. I've no idea what that looks like, but something's got to give.

I build up my pace as thoughts jostle for space in my mind but catching a glimpse of movement ahead, I slow to a trot and then a careful walk.

The movement comes from a scaly black and yellow tail that glides languidly from side to side as an animal, facing directly away from me, works away at something.

I stop in my tracks, staring at the beast that I've had near encounters with over the last fortnight, but never properly laid eyes upon.

As if sensing the cessation of my movement, the animal turns and regards me with a striking, reptilian stare.

The creature is large, most closely resembling a lizard, but the size of a crocodile. It has fat legs and an elongated head like a monitor. A long, forked tongue darts out between sharp teeth, tasting the air.

Its skin is mostly black with small yellow stripes and dots creating interesting patterns on its surface.

The lizard is huge, and no doubt powerful. In a dog fight, I know that I'd come off second best.

I take a step backwards and stop as the lizard moves as well. It lifts its right arm and waves it in the air in a large, slow circle.

'I'm sorry, I didn't mean to interrupt you,' I say instinctively.

I've no idea if the lizard's movement is a show of aggression or friendship, but something makes me take a step towards it.

The lizard bows its head gently.

I bow back and take another step forward. The creature turns and slowly walks away.

No longer feeling in danger, I follow the lizard carefully, as if in a trance.

The lizard crosses to a large tree and starts to climb until it's a few metres from the ground.

I stop at the base of the tree looking up in awe at this magnificent beast.

Realising that I can still see it, the lizard races up into the canopy of the tree, moving four times faster than before.

I follow its movement at first, but as it becomes obscured by the pine needles, its mottled skin blends with the foliage and I lose sight of it amongst the upper branches.

'Sorry again for disturbing you,' I call into the treetops. 'I don't mean you any harm.'

I cross to where the lizard had been standing and see a large mushroom, uprooted and half eaten.

I take a minute to soak in the experience and to let my adrenaline levels return to normal.

The lizard seems harmless, and if it's the same beast that has crossed my track before, then it's had plenty of opportunities to attack me. It's at least as large as me, and its claws are sharp as razors. However, instead of attacking, it fled up the tree, using its claws and slender body to ascend far more nimbly than I can.

It was a short, but magical moment. I decide that I'm not going to tell the others about the lizard. I'm not sure why. Possibly because of the growing schism with Declan, and because, in this new world, we need to share nearly everything with each other. This little thing I want to keep for myself, at least for the moment.

I return to camp and help with the ongoing construction work. Even despite the simmering feud, we all work well together. Everyone knows their own strengths and weaknesses and participates appropriately.

By dinnertime, we're all tired and conversation has largely petered out. Nat serves out a typical meal. The new types of mushrooms that excited us a few days ago already seem boring.

We each hunker over our plates, consuming the food for the nutritional benefits more than for the taste.

Tomohiro breaks the silence with his now familiar, lopsided, slushy drawl. 'So. Let's talk about sex, baby.'

Heads bob up around the circle to look at him.

'What?' he says. 'We need to talk about it sometime. What's the plan?'

'I think it's best if we wait until all of the volunteers are awake before we consider it,' I say.

'That's stupid,' Declan says. 'Nat and I are already doing it. Why wait?'

'Because we need to make sure we've built a safe society before we consider bringing new babies into the world!' Cassidy says in exasperation.

'Hey dudes, relax. They take nine months to grow, remember?' Jay says. 'We'll be sorted by then.'

'Hopefully,' Karim adds.

'Either way, we have to be very careful,' I say. 'Our genetic pool is going to be small enough as it is, once we each have a sexual partner, that's it. There can be no swapping and changing.'

'You're such a bore, Sarah,' Declan says. 'Live a little. You're taking all of the sexiness out of sex.'

Hearing him talk about sex makes my skin crawl.

'She's trying to create a sustainable society, thank you very much,' Newton says. 'If the two of you insist on having sex, then go for it,' he says, waving a dismissive hand at Declan and Nat. 'But the rest of us will wait until the others have woken up and partner off in a sensible manner.'

'Sorry for getting everyone worked up,' Tomohiro says.

'It's not your fault, Tomohiro,' Cassidy says. 'It needs to be discussed.'

Tomohiro looks at his lap, focussing on his limp left arm. 'It's just that the most productive thing I can offer to this society is strong and healthy offspring.'

'And so much more, Tomohiro,' I say. 'You'll make a great Dad, but there's no rush.' I move to sit beside him. 'You alright, mate?' I ask quietly.

'Fine enough,' he replies. 'It just hit me again the other day when you all went to the spring and I was stuck here.'

'I know, I'm sorry we left you.'

'Do you think you could take me?' he asks. 'To the spring?'

'There's a few spots that are really tricky with the chair…' I begin. 'But of course we can manage that,' I say hurriedly in response to his downcast expression.

He can't hide his smile. 'Really? That'd be so great.'

'Of course. I reckon with four of us we'd be able to manage you and the chair just fine.'

The following day Jay, Karim, Newton and I agree to take Tomohiro to the spring. We each load a backpack with dirty clothes to clean. We have no soap, but the clean water will hopefully freshen up the cotton garments.

I check in with Cassidy to make sure she's okay to be left behind with the other Alphas, but she assures me she'll be fine.

As I head off with the boys, we chat idly, and I get to know a bit more about Jay and Tomohiro. At only twenty four, Jay is one of the youngest amongst us. He worked at his family's butcher shop for most of his childhood but went to university to study physiotherapy. He wanted to join the R.E.Generation project because he always felt a great spirituality when his family took him onto country and he wanted to ensure that there was someone to care for country in this new world. He applied for the Project before he'd graduated from uni and was initially put on an extended overflow list before getting a late call up. He tells us that he was only two weeks into his first job as a physio before he had to quit to go under. He got even less training than the rest of us.

Tomohiro tells us that he's the eldest of the group. I'd assumed that he was in his late twenties, like most of us, but he's actually thirty-six. He worked as a fly-in fly-out miner straight out of high school, before heading to university to study geology. At university he learnt to climb and cut down trees with the Arborist Society. After graduating, he worked for a big mining company for a few years before getting out of the industry and taking an apprenticeship as a blacksmith.

'I'm so jealous of you guys for being able to climb the weather tree,' he says. 'Or any tree for that matter. There's a lot of things I miss.'

'I know what you mean,' Karim says. 'I mean, no I don't. Shit. Not in the same way. Sorry, mate. I can't imagine what this is like for you. And this is gonna seem really trivial compared to what you've lost, but man, I miss showers.'

'It's alright, mate,' Tomohiro says. 'I know what you mean. I miss feeling properly clean.'

'I miss my phone,' I say. 'Is it too much to ask for the internet to still exist?'

'I mean, are the satellites still up there?' Jay asks.

'Nah, mate,' Karim says. 'The cascade of space junk collisions would've smashed all the satellites to tiny smithereens.'

'Shame, I'd love to live stream some music,' Jay replies. 'Is that all gone? All the music?'

'Unless they put it somewhere safe, then I guess so?' Newton says.

'What do you miss, Newton?' I ask.

'Good food.' We all laugh. 'And comfortable furniture. It's okay for a few days when you're camping, but knowing I might never sit on a couch again, that's depressing.'

'I miss my bed,' Tomohiro says.

'I miss my parents,' Jay adds.

'My cat,' Karim says. 'Aslan. He died a few weeks before we went under. But I still miss him so much.'

'Man, what have we done?' I ask. 'Are we insane?'

'Probably,' Newton says with a smile. 'But that's what's gonna make this all so interesting.'

We continue our trek uphill, taking it in turns to push Tomohiro in the wheelchair. At the cliffs, two of us take Tomohiro up under our arms and the others lift the chair up the slope.

Tomohiro seems reasonably steady on his feet, stepping with his good leg and swinging his bad leg through to briefly take some of his weight. We transfer him back into the chair and twenty minutes later, we hear the burble of the small stream up ahead.

We follow the sound to where the spring water emerges from the rocks and begins its journey down the far side of the hill.

I take out my canteen, empty the dregs and fill it with spring water. I hand it to Tomohiro, who drinks deeply.

'That's delicious,' he says with a huge smile. 'Thanks so much, guys. This is really special, and I've had a wonderful morning.'

'Me too,' I say. 'I think this was exactly what I needed.'

Tomohiro appraises the spring thoughtfully. 'Has anyone tried diverting it?' he asks.

'What do you mean?' Karim asks.

'Well, it emerges here and flows down the hill away from our camp. But it's also downhill towards our camp, so if we can divert it a bit up here, we might be able to get it to flow closer to us.'

'Duh, that's so simple,' Jay says. 'Why didn't any of us think of that?'

'See, I told you that we need you, Tomohiro,' I say. 'You've got a wise old head on those shoulders.'

'Who are you calling old?' he says jokingly.

'You,' I say teasingly. 'I think I'll call you silverback from now on. Thirty-six. Gee, you're basically ancient.'

Our first attempt to redirect the water is unsuccessful. The spring emerges from a crack in a slab of rock, and we try to place some stones on the slab to force the water towards us. Most of the water trickles around the stones and the stones themselves keep sliding down the angled rockface.

Next, Karim, Jay and Tomohiro try and dig a channel from the small pool at the bottom of the rock, while Newton and I rinse and ring out the clothes. The water naturally overflows on the northern end of the pool, but by cutting a deep crevice into the hilltop towards the south, we will hopefully be able to create a new stream. We have no tools with us, but by using large sticks and their hands, the boys are able to create a short, stagnant off-shoot from the pool. The soil is soft and loamy, but large roots and rocks hinder their progress. The crest of the hill is a metre or so higher than the pool, so a much deeper channel will need to be dug to force the stream to the other side. We decide to return with metal tools, but their small progress shows there is some hope of making Tomohiro's idea work. They help Newton and I with the last of the clothes and we bundle the damp garments back into our bags.

My feet start to ache on the seven-kilometre hike back to camp. We spend a lot of our days on our feet, building, exploring and generally walking around. I take a moment to realise it's only a matter of time before all of the shoes we brought with us fall apart completely and we won't have anything at all to replace them.

We trudge back into camp, weary from the hike and needing to push Tomohiro most of the way.

Cassidy is mending a large gash in one of Newton's shirts that he damaged while climbing the weather tree.

It looks like Nat, Declan and Ayesha have finished the final touches on the second hut, meaning that we'll all be able to move in as soon as we want.

We inform the others of Tomohiro's plan and they are all excited at the idea of not needing to hike for fresh water.

'We could even build an aqueduct,' Declan says enthusiastically. 'If the Romans could do it, so can we.'

We agree that the slope from the spring to our camp should hopefully do most of the work for us, but Nat and Karim start to design a wooden aqueduct to complement it where needed.

'Now that we've finished the two new huts, we can wake the rest of the volunteers, yeah?' Ayesha asks.

'I was thinking the same thing,' I say.

'And with seven kilometres to cover, the more hands we have digging the channel and building the aqueduct, the better,' Karim adds.

We still have most of the afternoon, so we set about devising a plan for waking the other volunteers.

SATRIA

The word refugee has always meant 'other'. Other people coming to Satria's country. Other people who have no connection to Satria. Other people who may be illegal immigrants. That subconscious thought process, whirring away in the back of Satria's mind, has allowed him to sit back and watch their plight without feeling proper empathy. Without caring for those other people. Many of them passed through Indonesia, heading for Australia or other places of refuge, but many remained, and he had no interest in getting to know them. In fact, he would've preferred if they'd never arrived.

All of that has changed now. Refugee suddenly means 'me'. Refugee means 'us'. Satria finally understands that no one sets out to become a refugee. Just like him, none of the refugees who came before had 'refugee' in their life plan.

Satria stands in the prow of his boat, watching his island get smaller and smaller as he leaves for the final time, taking his wife and five children with him. Soon, he cannot see the island at all as it is obscured by the ocean. A fitting analogy for the way it has been slowly devoured by the rising seas over the past decades.

To be fair, they probably should've left a few years earlier – many of their neighbours did – but Satria held tight. Unfortunately, the sea was always going to win. This year, when the storms came, the sea arrived at their front door. After each storm, the sea would recede, but soon it will cover the island completely.

Now that the storm season has passed, Satria is taking his family to safety. First to Jakarta and then to Australia. There's no point in staying in Indonesia. There are so many other people displaced that all the higher ground has already been claimed, and sooner or later, most of the islands will be swallowed entirely.

Hopefully, when they arrive in Australia, the people there will show more compassion for Satria and his family than Satria did for the refugees arriving in his country.

The seas are kind to them as they sail to Jakarta. They have plenty of food and Satria is a capable fisherman, allowing them to eat fresh fish every night. The children are restless, restricted to the confines of their small fishing boat, but entertain themselves by playing silly games.

'Can you believe we'll never see our home again?' Satria's wife, Farah, asks him on the third day.

'We may never see our homeland again,' he replies. 'But we will see our home. We will make a new one, with space for the children to roam freely. Wherever they are will be home.'

She nods. 'Do you ever question if it was right for us to have children?'

Satria looks at his wife in confusion. 'What do you mean? We both love our children. I couldn't imagine a life without them.'

'I know. But was it fair to them to bring them into this world? Even the big countries are struggling with the war that the earth is waging against us. We think that we're heading to safety, but Australia is ravaged by bushfires and cyclones. Europe is experiencing its hottest summers and coldest winters. What sort of a world are we leaving our children?'

Perdana, their youngest, is trying to climb the mast. He has a huge smile on his face, and he laughs along with his siblings when he falls and lands on his bottom.

'Do you see that?' Satria asks. 'The naivety of youth that means they don't worry about those things like we do. That happiness and joy. That's the reason why I don't question bringing them into the world, because the world needs more of that. We need youth, exuberance and hope. Maybe Perdana will become the scientist that makes a breakthrough that changes all of this.'

Farah nods, hoping her husband is correct.

That night, Satria is woken as the boat rocks violently to and fro. He exits the sleeping quarters and clambers onto the deck. A fierce storm has arisen from nothing and is threatening to capsize the boat.

Satria hears a small voice call to him through the howling wind.

'Are we gonna be okay, Daddy?'

He turns to see Perdana stumbling across the deck towards him.

'Get back below deck,' Satria yells, but it is already too late.

A huge wave washes over the deck, knocking Satria off his feet. He tries to keep an eye on Perdana as he scrambles to grasp the siderail, but with sea water in his nose and mouth, he loses sight of him. Once Satria regains his feet, there is no sign of the boy.

Satria rushes towards the sleeping quarters, scanning the surrounding water for any sign of his son. He spots an arm, splashing amongst the whitewash before it disappears below the surface.

'Farah, on deck, quickly. Perdana is overboard!' Satria yells down through the hatch.

He doesn't wait to see if his wife has heard him, throwing himself over the side of the small vessel and powering through the water to the last place that he saw his son.

He dives under the surface, scanning the murky water for anything. The salt stings his eyes, but having grown up around the ocean, Satria has learned to keep them open underwater. He is thankful of that now as he sees a red smudge floating below him. Perdana's shirt billows around the lifeless body of his son.

Satria dives deeper, his lungs screaming for air. An adrenaline surge allows him to stay under the water for longer than ever before. He wraps an arm around Perdana's waist and kicks out hard for the surface. As they rise, he has a moment to marvel at the stillness of the water this far below the storm. As he reaches the surface, the chaos returns and it's all he can do to keep Perdana and himself above the water.

He swims on his side towards the boat, dragging Perdana with him. Farah is leaning over the edge of the boat calling to him, but he can't hear her words over the storm. Satria can feel his strength fading, but there is no way that he isn't getting back to the boat.

He touches the rough wood, grasping for a handhold and Farah pulls Perdana out of his grasp. She checks the boy over while Satria holds onto the edge of the rocking boat, trying to muster enough strength to clamber back onboard.

'He's alive,' Farah says emphatically.

Satria takes an enormous sigh of relief and Perdana splutters on the deck, coughing up the sea water that he inhaled.

Satria takes a few deep breaths and starts to pull himself up the side of the small boat. He is almost over the edge when he hears a loud crack. As he glances up, the mast comes crashing down, heading straight for him.

The mast hits Satria across the skull with a sickening crunch and he falls back into the water.

Farah rushes to the boat's edge. Unable to swim in such stormy waters, the relief that Farah felt at the return of her son disappears instantly. Fear and bodily anguish overwhelm her as she watches her husband's unconscious body float away into the blackness of the great ocean.

SARAH

CHAPTER 8

We need medical expertise to awaken any volunteers who are still alive inside their pods, and as such, Cassidy and I are put in charge of reanimating the volunteers from our original groups.

As none of the Gamma volunteers are yet awake, Declan is designated as the medico for their unit.

We're reasonably confident that the remaining three Alpha pods are bust, but Tomohiro volunteers for the grisly job of prying each one open and checking inside.

Karim agrees to help me with the Eta pods, while Newton and Ayesha are set to help dig out Epsilon pods from the landslide. Nat and Jay offer assistance to Declan, who will also need to unearth pods before he can awaken any volunteers. With any luck we could still find up to fifteen more volunteers. Even taking into account some damaged chemo-pods under the landslide, I'm hopeful of adding at least ten more people to our team.

Karim and I walk towards the Eta pods in silence. I've only returned a few times since we awoke from our own chemo-pods, and I'm not sure Karim has been back at all.

I show him Kyle's skeleton and Ang's pod, prised open in the treetops. We look for Mabel and Pam's chemo-pods but can see no sign of them, either in the trees or the surrounding forest.

We turn our attention to the two unopened chemo-pods lying on their sides. The large matte-grey incubators have tumbled together, pushed aside by some pine trees, but look otherwise intact. They each have an exposed port on the underside where they were ripped clear from the small geothermal power supply buried in the ground below the original R.E.Generation building.

The pods are caked in dirt and dust and have a scattering of rubble nestled between them from when the roof fell in.

We manoeuvre them as carefully as we can so that they are the right way up. We decide to open one at a time, starting with the one on the left that we assume to be Sam's pod. Karim finds the release mechanism and opens the lid slowly.

Sam lies in front of us, eyes closed and motionless. His face is calm and serene, with a bushy beard enveloping his jaw and long curly hair falling to his shoulders. The oxygen mask over his face appears to be having some difficulties sitting closely enough to his mouth, due to the mass of facial hair. Sam is muscly but thin and short statured, approximately the same height as me.

The wire mesh full of needles covers his body and the metal heart-pump rests over his chest. His left arm has shifted within the cradle that was supposed to keep him completely stationary in the chemo-pod. His arm is covered in many cuts where the hundreds of needles have shifted within his flesh.

I point the wounds out to Karim. 'You'll need to be ready to bandage that when his blood starts flowing. It's gonna be a bloody mess.'

He nods his understanding.

I flip open a compartment containing the override reactivation equipment. I need to input a code, but the machinery is completely drained of power. I know that there is a manual override ability, whereby I use a hand pump to distribute the reactivation chemical alongside CPR. It's incredibly hard to get the process perfect and we were told only to use this as a last resort.

During training, I only had a twenty percent success rate during manual reactivation simulations.

'It's out of juice. Do you think you can rig up power for it?' I ask Karim.

'I can try. Our chemo-pods have power, so I'll try and swap it out for one of them.'

I lift up one end of my pod and Karim carefully disconnects it from its underground power plant. We move it to the side and I'm somewhat surprised to see some of the original concrete slab intact underneath.

We close the lid of Sam's pod and drag it towards the slab. We prop one end slightly off the ground so that Karim can access the underside and he tinkers with the power port for a few minutes before carefully lowering it back down.

'It should be online now,' he says.

I pull the lever and the lid swings open. Inside, the displays show varying start up messages and loading screens.

I type in the access code and answer a series of questions ensuring that I want to begin the reactivation process. This finishes with needing to accept the *Terms and Conditions* which I don't bother reading before selecting *Confirm*.

Next, I grab a small lever and after a quick flick, the pod whirs into life.

As with Newton's wake up, Karim and I must simply wait as the chemical is diffused and the mechanical pump presses on Sam's chest, forcing viscous blood around his body.

After some time, the needles retract and Karim steps up to the pod, taking off his shirt. Blood starts to seep from some of the wounds in Sam's shoulder. The wounds on the rest of his arm do not bleed, indicating that the blood is not flowing there. Karim ties his shirt around Sam's shoulder, bandaging it as best he can.

We need to tie it quite tightly to prevent excess blood loss, but this further limits the spread of fresh blood and chemical to the lower parts of the limb.

Sam is breathing of his own accord and suddenly, the steady rhythm is replaced by a sharp intake of breath, followed by a moan.

'Are you alright, Sam?' I ask tentatively.

'Fuck, that hurts,' he says without opening his eyes.

'And so, your first word in the new world is an f-bomb,' Karim says.

'I guess that's true to fucking form,' Sam chuckles.

We grin.

'Are you alright?' I ask again.

'Fine enough. Just following procedure.'

He has finished stretching his toes and ankles and starts flexing his fingers. His right-hand closes into a fist a couple of times but his damaged left hand makes no movements.

'Give that a bit more time, I reckon. It still could come good,' I tell him.

He opens his eyes and appraises us as we lean over his pod. 'Sair, Karim. It's bloody good to see you.'

'You too, mate,' Karim says.

'You know,' Sam continues. 'My biggest worry was that I was gonna be the only one who woke up and I would be totally alone.' He looks past us and starts taking in the sights of the surrounding forest.

'Yeah, that was me for a bit,' I say. 'It wasn't nice at all. Bloody scary, if I'm being honest. But there are nine of us now, ten including you.'

'Neat trees. The pine smell is calming,' he says, inhaling deeply.

I look at the forest, remembering what it was like when I first woke up. The roof of pine needles and the lack of bird song. The sense of something completely new.

'It's amazing how quickly you get used to it,' I say.

'Absolutely,' Karim agrees.

As we prepare Meaghan's pod, we fill Sam in on the story so far. How long we've been awake. What we've been up to.

He sits up in his pod, nodding attentively. His damaged arm hangs limply at his side, so I quickly fashion a sling out of my own t-shirt to elevate it.

The blood from his shoulder is already seeping through the bandage fashioned from Karim's shirt. We will need Cassidy or Declan to look at that as soon as possible.

We move Karim's chemo-pod off its base and repeat the process of connecting Meaghan's pod to power. Once it's in place I instruct Karim to take Sam to Cassidy, as I turn to the pod's controls.

Air whistles into the pod as I raise the lid and thankfully, Meaghan appears injury free. I initiate the awakening process and wait patiently.

After five minutes, the mesh and heart-pump lift up and into the lid, the protective cover sliding smoothly across to lock them away. Another five minutes later, Meaghan opens her eyes and cautiously looks around.

I stand upright from my slouched position and smile at her. Meaghan had been my least favourite Eta volunteer in training, but in this place, I have to put all that behind me.

'Sarah,' she says curtly.

'Meaghan,' I reply.

'So, you woke up before me, then. Huh?'

'And a few others too,' I tell her.

Meaghan was always very competitive in training, and I get the feeling that she feels inferior for not being the first awake.

After completing her stretches, she sits up and considers the trees and scattered chemo-pods.

'Who's awake then?' she asks.

'Sam and Karim. A bunch of Alphas and two of the Epsilon volunteers.'

She nods and climbs out of her pod. Meaghan isn't classically beautiful, but she has a striking face and a figure to match. She was an elite athlete and competed at an Olympic games in the modern pentathlon. She also completed a law degree while raising a daughter of her own. Her daughter was twelve at the time we were put into chemopreservation. When I asked her if she felt sad about leaving her daughter, she told me that if her daughter was smart enough, she would become a R.E.Generation volunteer herself.

Meaghan ties her long, scarlet hair into a high ponytail and rolls up the sleeves of her jacket. 'Alright, what needs doing?' she asks.

I tell her that most of us have been awake for several weeks and that the others are also waking slumbering volunteers. Before we leave, she pulls a hunting rifle out from her pod and slings it over her shoulder.

'Are there any animals for hunting?' she asks as we walk through the forest.

'Not really. We've found two small snakes in trees that we've felled, and a few poisonous grubs. That's it, I think.' I fail to mention the large lizard that I'm still keeping secret.

'Shame.'

I nod.

We head towards where Cassidy, Newton and Ayesha should be working to dig out more Epsilon volunteers. When we arrive, I see a large pile of dirt and one freshly excavated chemo-pod. Newton, Ayesha, and Karim are digging into the hill with trenching tools looking for more pods.

'How's everything going?' I ask.

Newton shakes his head at me, looking dispirited. 'Not good. Not good at all. It's slow work with these tools and the pod that we did find is burst.'

'Newton and Cass think that the weight of the soil has forced the pods open. Particularly the deeply buried ones,' Ayesha adds.

'Even if they weren't damaged in the initial landslide, the pressure from the soil over the centuries would've been slowly crushing them,' Newton says.

'Shit,' I say. There's not much more worth saying.

'I'm Meaghan, by the way,' Meaghan says, approaching Newton and Ayesha. They shake hands and Newton gives Meaghan a broad smile. I feel a stab of jealousy but try to quash the feeling. I've got no ownership of Newton.

'I'm so happy to see you, Meaghan,' Karim says, giving her a hug that she doesn't reciprocate. 'It will be refreshing having some new faces around.'

'How can we help?' Meaghan asks, getting straight back to business.

'Maybe go check on how the other Alphas are doing,' Ayesha suggests.

'Hopefully, they're having more luck with the Epsilon pods,' Karim adds.

'I assume Cass is looking at Sam's arm?' I ask.

'Yeah, she took him back to camp,' Newton replies.

'We'll catch up with you later then. Good luck,' I say, with an encouraging smile for Newton.

He briefly smiles back, but his overall body language is dejected. It's hard to see him so gloomy, but realising that you might not see any of the friends that you trained with must be incredibly sobering.

Meaghan and I climb up the short rise created by the landslide and descend the other side towards another excavation site. I'm encouraged to see Declan hunched over an open chemo-pod that appears to have someone inside. Jay and Nat are almost finished scooping out soil from around another pod that also looks intact.

It appears that neither pod was buried deeply, and hopefully that bodes well for the rest of the buried Epsilon pods.

As we approach, Declan starts talking to the person ensconced in the pod. 'Take it slow. No rush. There's no danger here. I'll look after you.'

A high-pitched, slightly irritating female voice replies. 'Thanks. You're Declan, right?'

'That's right. Do we know each other?' he asks.

'My dad was one of your lecturers at uni. When he heard you'd joined the R.E.Generation Project he encouraged me to join too. He thinks very highly of you.'

'Thought,' Declan says. 'He thought very highly of me. Unfortunately, your father died thousands of years ago, like the rest of humanity.'

'Yes of course. Of course. It feels like I only just went to sleep. Where did the time go?'

'You get used to it,' Declan says, offering her his hand. 'What's your name?'

Natalie looks over suspiciously, no doubt surprised by Declan's familiarity and noticing, like I do, that he's asked for her name, but not her profession.

'Tiffany,' she says, followed by a laugh. I can't tell if it's due to nervousness or entirely put on.

Tiffany is short and stocky. She has peroxide blonde hair and piercing eyes. One of the things I noticed during training was that almost all the volunteers were attractive. It would seem that in searching for 'perfect specimens' to seed our new civilisation, the recruiters for the R.E.Generation Project let an unconscious bias effect their decision making. Because, of course, we all know that attractiveness is paramount when it comes to being a skilled or successful person.

'And what skills can you bring to the team?' Nat asks curtly.

'I have ten years of experience as a physiotherapist, and I'm also very good at baking,' Tiffany replies.

'So, nothing particularly useful,' Nat mutters, turning back to her digging.

'What was that?' Tiffany says, approaching Nat aggressively.

'Nothing,' Nat grumbles, not looking up.

'You said that I'm not going to be helpful. Why's that, babe?' Tiffany says, probing.

Nat looks Tiffany straight in the eye and stands to her full height. She practically dwarfs Tiffany and she looks ready to punch her with all the might of her muscled shoulders.

'Jay here is a physio, so your skillset is not particularly needed and good luck finding anything to bake with. Sorry to cut you down, *babe*, but you're basically redundant.'

'Enough,' Declan says, cutting the confrontation short. 'We need more people so you're very much welcome, Tiffany. Nat, you're still my second in charge, so cool your jets.'

For once, I'm impressed with Declan's handling of the situation as both ladies seem sedated.

I thought waking more people would make life easier. But more people means more personalities, and more opportunities for friction.

'Hey, Dec, this pod's ready now,' Jay says, gesturing to the pod he and Nat had been working on.

Meaghan and I take the trenching tools from Nat and Jay as Declan opens the second pod and starts the now-familiar process of waking the volunteer inside.

Jay grabs a large stick and tries to help by digging it into the soft ground looking for more pods. Nat stands protectively over Declan's shoulder and Tiffany sits down to watch proceedings.

Meaghan and I dig deep trenches into the side of the slope, hunched over to use the small tools. The trenches run directly from the two unearthed pods, and I expect to find the adjacent chemo-pods reasonably quickly, but we both strike nothing.

'Hey, Tiffany, where were the other pods when you were in the lab?' I ask.

'Right there,' she says pointing to where Meaghan and I have dug.

I hear a clunk from behind me and turn around to see Jay standing proudly, leaning on his stick.

'Found one,' he says with a smile.

He is more than ten metres away, slightly downhill from my position. I can only assume the landslide must've carried the other pods with it.

We start to dig the pod out, but it soon becomes apparent that the airlock has been busted open.

We search deeper into the hill and eventually find another pod, which is similarly crushed and lifeless.

We're all feeling equally dejected when Declan approaches us with another beautiful woman on his arm.

'Everyone, meet Poh. Poh, meet everyone,' he says light-heartedly.

'Hello,' Poh says sheepishly.

'I think she's gonna be a big help. She's a structural engineer,' Declan announces.

Clever and pretty.

Poh has jet black hair and a tanned complexion. She has high cheekbones, fierce eyebrows and a rounded face with a gorgeous smile.

We make our introductions and decide to head back to camp to check on the others.

We meet up with the newly awakened Sam, who is with Cassidy and Tomohiro. Sam shakes Jay and Declan's hands and introduces himself.

'The name's Sam but everyone calls me Smithy,' Sam says warmly

'Nice to have you on board, Smithy,' Declan says sternly.

'How's your arm?' I ask.

Cassidy shakes her head behind Smithy's back.

'My shoulder hurts like a bitch, and I can't really feel the rest of it, but we'll see,' Smithy says optimistically.

'At least it's your left arm,' I say, trying to match his optimism.

'Except that I'm left-handed, Sarah,' he says.

I cringe and shake my head. 'Bugger. Sorry, mate.'

'It is what it is,' he says with a shrug.

Newton, Ayesha and Karim re-join us. No other volunteers follow, and I look around the group, regarding the four new additions.

It's hard not to feel dispirited considering the high expectations I set at the start of the day, but four more volunteers takes our total number to thirteen and is definitely better than nothing.

'I think a toast is in order,' Tomohiro says, trying to lift the mood. 'Everyone, grab a glass.'

Ayesha hands out metal cups filled with water.

'To R.E.Generation unit three-six-one-five-four-three. May we live long lives and make the most of this second chance.'

'To second chances,' most people echo.

Some people try to repeat the unit number with little success.

'You made that number up,' I say punching Tomohiro on the shoulder.

'Guilty,' he says with a cheeky grin.

I'm just glad to see him smiling. He seems to have taken to Smithy, and they spend the evening bonding over their injuries. Tomohiro takes great delight in telling Smithy all the things he's going to struggle to do one handed.

'Getting dressed is the worst,' he says.

'I can imagine,' Smithy says sombrely.

'And did you say your left hand is your dominant?' Tomohiro asks.

'Yup.'

'Then you're even gonna have difficulty wanking. Sorry mate,' Tomohiro says, unable to keep a huge, lopsided grin off his face.

'I bet you've been doing plenty of that, stuck in your chair, you legless twat,' Smithy retaliates with an equally big smile on his face.

I'll never understand the crassness of men, but at times, it does seem to help them get through adversity, so it can't be all bad.

Tiffany sits down beside me and gestures towards Tomohiro. 'Hey, babe. What happened to him?' she asks.

'Neurotoxin, we think, from a poisonous grub. Paralysed his entire left side.'

'Can he walk?' she asks.

'A little bit. But he spends all his time in the wheelchair now,' I respond, appraising her carefully and wondering where this conversation is headed.

'But he'll never get better if he doesn't try and re-learn to walk. Who's the other physio again?'

I point to Jay. 'Jay, come over here a minute,' I call.

He approaches us with a casual smile and a questioning look. 'Yes?'

'Why haven't you done any rehab with Tomohiro?' Tiffany asks.

'Well, I...he seemed to be managing in his chair,' Jay mumbles.

'But surely your first thought was how can we make this better for him? And he'll be losing muscle mass in his good leg if he just stays in the chair.'

'I did think of that. But I didn't know where to start. I don't really have any experience with people like him... Well, I don't really have much experience at all...' he trails off.

'Right. That's it, come with me,' Tiffany says taking control and marching over to Tomohiro.

She forces him to get out of the wheelchair and completes a

quick assessment of his abilities. She asks Jay lots of questions, to which he knows all the answers. As they progress, he seems to grow in confidence, realising he does know what he's doing.

They spend almost an hour devising a rehab program with Tomohiro, focussing on re-training him to stand independently and walk. Tomohiro makes little progress in just the single session, but he seems appreciative of Tiffany's hope and enthusiasm.

Without any more rain, the mushroom supply starts to dry up and we must search farther to get any food to supplement our dried stores. Smithy tries as best he can to help out around camp, but as he is still getting used to having one arm as a dead weight, he is ungainly and is best suited to simple tasks like picking mushrooms and carrying water.

A few days later, we walk to the spring together to fill several canteens that we carry in our rucksacks. Smithy's bad arm is still in a sling, and as he moves it to take off his pack, a noxious smell wafts over to me.

'Hey, mate, how's your arm?' I ask him.

'You know, just not really doing anything. It's fine,' he says casually.

'Can I take a look?' I ask.

He turns away, protecting his arm from me. 'I'd prefer it if you didn't. It's a bit ugly you know. All cut up.'

'That's why I want to see it,' I tell him. 'It's my job, remember.'

'I know. It's just, it really is fine, you know. It's not bothering me at all.'

'Let me look at it,' I say sternly.

He has been wrapping it in fresh cloth each day, but otherwise hasn't asked for any medical assistance.

As I peel off the makeshift bandage, the smell gets stronger and I have to force myself not to gag.

Smithy's arm is horribly discoloured in a mixture of greens and purples, and it appears to be partially swollen along its entire length. The smell reminds me of rotten eggs, and I cover my mouth and nose to protect myself.

I stand and look at Smithy sadly. 'It's gonna have to go, mate.'

'Nah, it's fine. It's just bad bruising. It'll get better,' he argues. I can tell that he doesn't believe what he's saying, but he's hoping above all hope that there's a way to save his arm.

'I'm sorry, Smithy, but it's dead. No chemical made it past your shoulder, so your arm is literally rotting. If we don't cut it off, you risk it spreading infection and necrosis to the rest of your body.'

Tears start forming in his eyes and I place a hand on his shoulder tenderly.

'It's my arm, Sarah. My good arm. What's a stonemason without his hammer arm, hey? Fuck.'

'I'm so sorry.'

We stay there for a few minutes while Smithy lets reality sink in. I rebandage the arm, more to keep the smell in than provide any benefit to the tissue itself.

We collect the water, pack our bags and hike back to camp.

Later that evening, Cassidy and Declan work together to surgically remove the dead limb. They work well as a team, probably because Cassidy lets Declan take the lead and assists him ably without pushing his buttons.

The distinction between flesh that was re-awoken by the chemical and dead flesh that never got the reactivation fluid is quite clear, leading to a relatively simple operation.

They remove a few millimetres of healthy tissue to ensure

that the resultant wound is healthy and cauterize the wound with implements heated over the campfire.

Smithy has no feeling in the dead limb, but because they're working on live flesh, we have to give him a large dose of painkillers and a stick to bite down on.

After the operation he is still a bit high from the drugs and flaps his short stump around proudly.

'This is the sign of a true hero,' he says enthusiastically. 'The enemy may have shot me, but I lived to tell the tale. Three cheers for me.'

Later that night, I hear him sobbing in his sleep and the next morning, once the drugs have worn off, he has to deal with a constant, throbbing pain.

Tomohiro tries hard to keep his spirits up, but I also file away a note to keep checking in with Smithy. It's a big loss that he's gonna have to keep dealing with for the rest of his life.

We spend the next few weeks hiking up to the spring and slowly rediverting the stream towards the camp. We'd hoped that once we had the stream on our side of the rise, that it would flow towards our camp naturally. Once we've dug the initial channel, however, we realise that the slope of the hill takes the stream obliquely past our camp. This decreases the hike to about four kilometres, but we've all got excited about the prospect of running water in camp, so we agree to try and force its path more directly.

Poh devises a plan to build a series of dams and wooden aqueducts. Its heavy work, felling trees, splitting them in half and hollowing them out, but everyone puts in and we make steady progress.

The other complication is that the spring only produces a small flow of water, so each time a ditch or dam needs to fill it takes a long time before it's ready to overflow into the aqueduct.

Thirteen days into the build, I am inspecting a part of the aqueduct on my own when the sound of a gunshot rings out nearby and echoes menacingly around the forest.

ADRIANA

Adriana is a perennial optimist and can't help feeling that she's different. Unlike her friends, she will be fine. The tests will give them the answer they need, and they will be able to have the children they so dearly want.

The doctor calls Adriana and her husband into his room.

'Adrian and Adriana?' the doctor asks with an eyebrow raised quizzically.

'You know it's true love when you choose to marry someone with the same name as you, don't you?' Adriana says with a twinkle in her eye.

'I guess that's true,' the doctor says in a slow drawl.

'So, what do the test results say, Doc?' Adrian asks.

'I'm sorry to say that the results aren't good. You are infertile and having children will therefore be almost impossible for you,' The Doctor says bluntly.

Those words knock all the wind out of Adriana, but she can't help clinging to hope. '*Almost* impossible' is what he said.

'It's me, isn't it?' Adrian says.

'Well, actually, it's both of you.'

'How is that possible?' Adriana asks in anguish. 'We've been so good. We've been avoiding plastic, avoiding chemicals. Only using natural products.'

'*The Blight* has affected most people, and we still don't know exactly what is causing it. But if it makes you feel any better, it probably isn't even your fault. These flaws were probably laid down while you were inside your mother's womb. Or even when she was inside her mother's womb. The egg that would one day create you was already in existence. Your grandma had some Chinese takeout, the hot food absorbed plastic particles and chemicals. After she consumed these, they were absorbed by her body, flowed through her blood and through the placenta to your mother. Once in your mother's body one little chemical molecule said, '*hey, that's an interesting egg*,' and just like that you're now infertile. Now that's obviously over simplified and just one possibility. But unfortunately, the stark truth is that you're both infertile.'

'Can we use a donor egg?' Adriana asks.

'Of course we *could*. We have plenty of ova and spermatozoa frozen from the last hundred years, from before *The Blight*, but I'm sorry, Adriana, your uterus is incompatible with pregnancy.'

'How can you know that without even trying? You said that it would be almost impossible for us to have children. Almost impossible is not impossible.'

'We have years of data, test results, and the experiences of thousands of other women like you that tell us that this will not end well.'

'But I'm different. I want to try,' Adriana says.

Adrian squeezes her hand supportively.

'Naturally, you can try, but I can't recommend it. I need you to remember the incredibly low chance of success; we're talking one in a billion. And the chances of complications, that put your health and that of the foetus at risk, are much higher than that.'

'Doc, we're not giving up on this,' Adrian says. 'So you'd better get to work.'

Adrian rests his ear against Adriana's swollen belly.

'Can you hear anything?' she asks.

'Not really,' he responds. 'Just regular body sounds. Churning and what not.'

'But our baby is in there,' Adriana says with a huge grin on her face. 'Our baby. And we only have to wait four more months to meet her.'

'I'm so glad we got to prove that doctor wrong,' Adrian says. 'When should we start thinking about names?'

'Why not now?' she replies. 'But I think I need to go for a walk. It all feels a bit uncomfortable at the moment.'

They grab their coats and start putting their shoes on. Tilly, their border terrier, races up to them, her tail wagging furiously.

'You know where we're going don't you, you clever girl,' Adriana says.

'Alright then, you can come too,' Adrian says grabbing her lead.

As they walk down the stairs of their apartment building Adriana grabs her side and gasps.

'Are you alright?' Adrian says worriedly.

'Yeah, I'm fine,' Adriana manages to say, breathing through the pain. 'It's just a stabbing pain, but it goes away.'

'Well, let me know if it gets any worse, okay? I can call the Doc and get you booked in for a check-up. We can't be too careful.'

She nods.

They walk along the foreshore, enjoying the fresh sea breeze and the late winter sunshine. People stare at Adriana as she walks past. She's just begun to really show, and people aren't used to seeing pregnant women anymore.

It's one of the few nice days that they have had in the last month. All winter, the weather has been disturbed, fluctuating between cold snaps and heat waves as if months had passed rather than just a few days.

'I think I need to go to the bathroom,' Adriana says.

'Alright, there's a public toilet just up here, I think.'

Adriana disappears inside the loo and Adrian waits outside with Tilly. Someone has left their labradoodle tied up to a pole and Tilly and the other dog enjoy sniffing each other and making their introductions.

After five minutes, Adrian starts to get a bit worried, but Adriana has been going to the toilet more and more since she has been pregnant and she seems to be taking longer each time as well. After a further five minutes, however, his anxiety is peaking, and he waves down a woman who is passing by.

'Excuse me, could you go and check on my wife for me? She's in the toilets and she's taking a long time.'

'It's probably nothing, mate,' the lady says, turning to walk away.

'I know, but she's pregnant and I'm worried. Could you please just check?'

The woman turns around, intrigued at the thought of a pregnancy. 'How did you guys manage it? To get pregnant?' she asks.

'Does that really matter?' Adrian says, exasperated. 'Can you just check on my wife, please?'

'Okay, but afterwards you'll have to tell me which doctor you've been seeing.'

She disappears inside the toilet block and Adrian shakes his head. He's still not used to Adriana being a borderline celebrity just because she's pregnant.

The woman rushes out of the toilets and waves to Adrian. 'You'd better come in. She's not responding to me.'

Adrian ties Tilly onto the pole and hurriedly follows the lady inside. He calls to Adriana, but there is no response. One of the cubicles is locked and peering under the door, he recognises Adriana's shoes. The underwear around her ankles has a large blood stain on the lining.

'Adriana, are you alright?' he calls. 'It's me, Adrian. Whatever has happened, it's going to be fine. We'll work it out together. Just let me in.'

She doesn't respond and he kicks at the door, hoping to force his way in. The door doesn't budge, and pain radiates up his leg as he jars his ankle.

The lady moves into the adjacent cubicle, stands on the toilet and peers over the side. 'Oh my god,' she yells. 'We need to get in there. She's unconscious.'

Adrian tries to ram the door with his shoulder while the woman calls an ambulance. The whole line of cubicles vibrates but Adriana's cubicle remains locked. Adrian fumbles to find a credit card and manages to jimmy the lock open.

He rushes into the cubicle and falls to his knees in front of his wife. Adriana is slumped to the side, and the toilet bowl is full of blood. He checks her pulse but can't find anything and she doesn't appear to be breathing. He pulls up her pants to give her some modesty and drags her out onto the bathroom floor.

He starts CPR, his first aid training coming back to him. He wants to sob and collapse, but he manages to keep it together to give his wife a fighting chance at survival.

'See if you can find a defibrillator,' he calls to the woman.

She is as pale as a sheet, but she nods and stumbles out of the bathroom.

The lady returns with a red box and Adrian hooks the machine up to Adriana. He does everything that he can, but when the paramedics arrive, they tell him that Adriana is not coming back.

He curls up into a ball on the filthy bathroom floor and lets his emotions take over. He sobs and sobs. One of the paramedics sits with Adrian and tries to comfort him, but their platitudes are swatted away by his grief. The lady brings Tilly to him, and she snuggles up next to Adrian, feeling her owner's pain. He holds her close and she licks salty tears off his face.

The following day, he doesn't leave the couch and Tilly doesn't leave his side. The empathy of the little dog is seemingly boundless. He receives a call from Adriana's doctor who informs him that as the foetus continued to grow, Adriana's uterus stopped growing with it. The fibrous tissue was unable to stretch and eventually the foetus ruptured through the uterine wall, causing catastrophic internal bleeding.

Adrian is numb as the doctor explains the news, and one thought goes round and around in his head. *They should've listened to the experts.*

SARAH

CHAPTER 9

I rush towards the sound of the gunshot, hoping that no one is in danger and find Meaghan standing over the body of a large lizard. I can't tell for sure, but I think it's the same beast I met when I was running on my own.

'You killed it,' I say emotionally, unsure why I feel so protective of the beast.

'Yeah, it's meat, isn't it? Plus, it was acting aggressive,' Meaghan says bluntly. 'This is gonna make the best meal we've had since we've been here.'

She's right, of course, we need new food. But it still hurts that she's killed this beautiful creature. I also struggle to believe that it was being aggressive. It was so timid when I met it.

'You'd better go find Jay, then,' I say calmly. 'He'll know how we should butcher it. And maybe grab Tomohiro's chair. I'll stay and look after it.'

She nods and heads off towards camp. I sit beside the lizard and place a hand on its back in apology. I notice a large mound a few metres behind the lizard and cross over to it. I place my hand on the raised earth, which is warm to the touch. I dig in carefully with my hands and find a large, pale egg. I realise why the beast wasn't hiding from Meaghan. It was acting aggressively because it was trying to defend its nest.

Acting on instinct, and knowing that the others will be returning soon, I dig out three eggs and try to patch the mound back together so that it looks undisturbed.

I fold my top up in front of me and am able to carry the three eggs in my makeshift pouch.

I walk in a large circle around the camp before curling back towards it. I manage to avoid meeting any of the others and arrive close to the toilet and the original triangle huts. While many of the others have moved into the larger huts, Cassidy and I remain in our original hut, preferring to sleep on the forest floor, rather than on the rough-hewn timber flooring.

A few people are mingling around camp, but I manage to sneak into the hut without anyone seeing. I lay the eggs gently on the floor and push some pine needles around them.

Cassidy and I try to keep a small fire burning inside our hut to avoid having to relight it each night. There are still some hot embers in the pit so, using a blanket to protect my hands, I take two large stones from the side of the fire and put them with the eggs. Then I cover the eggs and stones with the blanket, hoping that the warm stone will provide enough heat for the eggs to survive.

I stoke up the fire and make sure there are more stones warming, before sneaking out of the hut, into the forest and retracing my steps back to the dead beast.

Meaghan has returned with Declan, Jay and several of the others. Newton, Ayesha, Meaghan and Poh are appraising the lizard.

'Where did you go?' Meaghan asks me when I re-emerge.

'I thought I saw another creature, so I followed it,' I lie.

'Did you get it?' Declan asks.

I shake my head.

'Shame.'

‘Is that an egg mound?’ Newton asks, pointing it out to the others.

Meaghan digs into the mound without hesitation and draws out a large egg. ‘Bacon and eggs for dinner tomorrow night?’ she says triumphantly.

‘Well, steak and eggs,’ Jay says, correcting her.

Everyone is jubilant at the thought of a good meal. I smile along with the others, pretending to be thrilled while hiding my hands behind my back to conceal the dirt under my fingernails.

We load the lizard into Tomohiro’s chair and Meaghan, Poh and Ayesha dig out three more eggs from the mound.

We make our way back to camp with the bounty and everyone chats merrily.

‘It’s huge, isn’t it?’ Ayesha says.

‘I reckon it could do some damage with those claws,’ Meaghan adds.

‘If it has eggs then there must be another one around. We’ll need to be careful,’ Jay comments.

‘I don’t think it’s dangerous,’ I say instinctively.

Newton turns to me inquisitively. ‘What makes you say that?’

‘Well, there’s no other animals for it to attack, so it must be a plant eater, or a mushroom eater.’

‘Maybe,’ Newton ponders.

Once we’re all back at camp, Jay butchers the large carcass carefully while Newton takes copious notes. He confirms from the dentition that the lizard is a herbivore and insectivore. Tomohiro spends that evening and the next day going through the procedure with the meat. Wrist, lips, spit, swallow. Luckily, he shows no side effects.

No one else seems bothered that we’ve killed a harmless animal, but I can’t shake the feeling of guilt in my stomach. Only the thought of the three eggs that I’ve squirrelled away makes me feel slightly better.

Once it's evening, I have to show the eggs that I am incubating to Cassidy. She is shocked at first, but soon agrees to keep my secret and to help keep them warm.

'But what are you gonna do if they hatch?' she asks.

'Raise them,' I reply simply.

'Aren't you scared of them? You saw how big they get.'

'Cows are big, people aren't scared of them,' I say defensively. 'Besides, they're not dangerous.'

'How do you know?' she asks.

I decide that I have to tell her the truth. 'I met one. It was far more scared of me than I was of it.'

'When?'

'About two weeks back. It was amazing, they have the most remarkable ability to camouflage in the treetops. They're hiders, not fighters.'

'I hope you know what you're doing,' Cassidy says.

I choose not to reply, roll over and wait for sleep to find me.

The following night, everyone gathers around the central fireplace as Nat cooks the meat and eggs into a dinner. She breaks one of the eggs into a large pot, but instead of finding an egg white and yolk, a small, largely formed lizard falls into the pot amongst some sticky goo.

I turn away in revulsion and I hear someone else gag.

'So much for steak and eggs,' Jay says.

'Oh well, it just means we have more meat,' Declan says.

'That's gross, you can't eat that,' Ayesha says, holding a hand to her mouth.

'Of course I can. Look.' Declan fishes the foetus out of the pot and skewers it on a long, thin stick. He cooks it over the flames, rotating it regularly.

Meanwhile, Nat adds cubes of lizard flesh and mushrooms to another large pot and fries them up.

The smell of cooking meat makes my mouth water, and when Nat hands around plates of the juicy meat I dig in ravenously, almost forgetting my guilt. It tastes surprisingly like chicken, with a flaky texture. It's the most flavourful thing we've eaten since we've been awake, and everyone goes back for seconds.

Declan bites into his skewered lizard and hands it around the circle. All the males try it, along with Nat, Meaghan and Cassidy. By all reports it tastes much like the adult lizard, but with a slightly different texture.

As we get up the following morning, we hear the sound of retching from across camp. Cassidy and I rush to investigate what's happening and find Ayesha doubled over next to a large pine. A small pile of vomit rests against the base of the tree and has also splattered onto her shoes.

'Ayesha, what can we do to help?' I ask.

'Nothing,' she replies. 'I'm feeling better already.'

'Do you think it was the meat?' Cassidy asks.

Ayesha shakes her head. 'I felt queasy yesterday morning too, before we'd eaten it.'

'Well, let's hope so,' I say. 'Otherwise, we're all fucked.'

Later in the morning, Nat complains of feeling ill as well which heightens my anxiety. It's possible that the meat didn't affect Tomohiro, but for some reason, the girls are reacting to it. Ayesha and Nat sit out the morning's work, however, Nat recovers enough to make us lunch. We're all tempted by the delicious meat, but decide not to risk it until we know more about Ayesha and Nat's condition.

Now that the new volunteers have been with us for a little while, we re-contest the election. I have spent the last week trying to build rapport with them but it is a foregone thing, as Declan has pulled Tiffany and Poh close as de facto Alphas. He wins the vote seven to six, and everything continues as before. I am pleased, however, that both Smithy and Meaghan vote for me.

I'm somewhat surprised that Meaghan didn't nominate herself to be leader. As a former Olympian, she is driven and competitive, but for some unknown reason, she threw her support behind me the minute we told the new guys about the vote.

We try to preserve slices of the lizard meat by smoking it over the fire. We store the rest of the carcass in the original Alpha hut and hope that it doesn't rot before we get the opportunity to make the most of it.

I return regularly to my own hut to check on the eggs, change over the warming stones and keep the fire burning.

The next morning Ayesha is sick once more and is joined by Nat, who throws up next to the Shit House.

Everyone gathers around, but no one else is showing any signs of illness.

'They couldn't be pregnant, could they?' I ask the group in general.

'Don't be ridiculous, Sarah,' Declan says.

'I'm not,' I say stubbornly. 'Morning sickness is a common symptom of pregnancy.'

'But Ayesha couldn't be pregnant, she hasn't had sex. Isn't that right, Ayesha?' Declan probes.

Ayesha glances at Declan nervously. 'Yeah, of course. I haven't. So, I couldn't be.' She stares at the ground, hating being the centre of attention.

'That's that then,' Declan says decisively. 'Must just be a bug or something. We'll have to monitor everyone, but we can keep eating the meat.'

'I'm not sure that's a good idea. We don't know that it's completely safe,' I say.

'Of course we do. We all ate it, what, thirty six hours ago? Most of us are totally fine,' Declan retaliates.

'Most of us, but not Nat and Ayesha,' I say irritably.

'I'm the leader, I say it's fine, end of story,' Declan says and storms off.

Tomohiro pulls me aside and whispers to me conspiratorially. 'You really shouldn't confront him like that. He's volatile.'

'Someone needs to stand up to him when he's being a prick. I think I can handle him.'

'I hope so,' he says, the tone of his voice sending a cold shiver along the length of my spine.

I leave the group to organise breakfast and head to the toilet. Once I'm finished, I push the fabric flap aside and am confronted by Declan who forces his way in, and blocks the exit. There's no way I can get past him.

There's hardly room for one person in the small building, let alone both of us. I can smell his body odour and bad breath. As it mixes with the smells wafting up from the drop pit, I almost hurl.

'We're going to have sex now,' he says calmly.

'Firstly, if I was going to have sex with anyone, it wouldn't be in a toilet,' I reply. 'Secondly, if everyone else dies and it's just me and you left, I still won't have sex with you. So please let me past.' My brain is whirring as I try to maintain my composure and stay in control of the situation.

'I don't think you understand,' Declan says, leaning towards me menacingly. 'It wasn't an invitation. It was a demand. You need to learn to stop contradicting me in front of the others. From now on, each time you do, we're going to have sex.'

I've never liked Declan, but this cold, calculating, disgusting man is something else altogether. Amongst the group, he uses a fake façade, but lying underneath is the real Declan. The Declan that has terrified Ayesha and Tomohiro for the last month.

I feel bile rising in my throat, but I force it down as I struggle to remain calm and think logically.

He's started groping my breasts aggressively, and I can see only one way out of the situation.

'Okay, let's get this over with then,' I tell him.

He seems a little surprised, suspicious even, but as I stroke his chest, he starts to relax. He half closes his eyes, no doubt imagining someone that he actually likes touching him in this way.

With all the speed and force that I can muster I drive my knee towards his crotch. He must have half-expected this as he twists just in time and I only land a glancing blow.

Declan grunts in pain and drops to one knee, allowing me to force my way past him and out of the Shit House. He stumbles after me, holding his crotch. I glance back in satisfaction as I see tears rolling down his face.

The campfire is out of sight from the toilet and as I round the corner of a hut, I slow down and march towards the others with purpose.

'This has got to stop!' I say, interrupting conversations. Everyone turns to look at me.

'What needs to stop?' Smithy asks patiently.

'Declan and his intimidatory behaviours. He just tried to rape me, so that I'd "stop embarrassing him in front of everyone else."'

Karim is halfway across the circle looking like he is going to kill Declan, before I manage to get an arm across his chest, to stop him doing something that will make this whole thing worse.

'No, I didn't,' Declan says calmly, walking towards the circle. His composure is amazing. I notice the slightest of limps but the others probably wouldn't. His eyes are slightly glassier than usual, but there's no sign of the tears he cried only seconds ago.

'I was just getting my water,' he says holding up his canteen.

'Fuck you, Declan,' I say aggressively. 'I want a group vote. Declan tried to rape me and there needs to be consequences.'

'He says that he didn't rape you,' Nat says. 'Why should we believe you?'

'You've had it in for me from day one,' Declan says quietly. 'Maybe you're the one who should face consequences for constantly trying to undermine your democratically elected leader.'

I look around the group. Even some of my friends are finding it hard to meet my gaze. 'Does anyone believe me?' I ask.

'I believe you, Sarah,' Cassidy says.

'Has he tried this with anyone else?' I ask. I try to catch Ayesha's eye without making it obvious. She has shut down, and stares at her feet. I really need her to speak up, but I'm not going to force her to.

No one says anything.

'You're the lawyer, Meaghan, what do we do?' Cassidy asks.

Twelve pairs of eyes swivel towards Meaghan who smiles. This is the moment she's wanted since she woke up. Her chance to be the most important.

'Obviously we can't have a proper trial, as everyone knows the accuser and the defendant,' she begins. 'There is no way that we could have an unbiased judge, or jury for that matter, so the only thing to do is to hear the complaint in full and have a group vote.'

Meaghan makes me explain my 'alleged' attempted assault in full detail. She then questions Declan who provides little additional information, citing that he was humming a tune to himself on the way back from his cabin when he heard me exclaiming to the group and came to defend his honour. What a prick.

'Any further evidence?' Meaghan asks.

I glance at Ayesha again, but there is silence.

'Ok then,' Meghan says. 'Let's vote as to whether Declan should face any consequences for this alleged, attempted assault. Which is, in essence, whether you think he is guilty or not. Votes for not guilty?'

Declan and I are ineligible to vote, meaning I need at least six of the remaining eleven votes. My hopes rise as I look around the circle. Initially only Jay, Nat and Tiffany raise their hands in the air. Then I notice that Ayesha has also lifted her hand slightly off her lap, and after a stern look from Nat, she raises it higher. Poh belatedly raises her hand to join the others.

Still, five votes are not a majority. Tomohiro hasn't raised his hand, but as I try to catch his eye, he avoids my gaze.

'Votes for Declan to face unspecified consequences, which are to be determined by another vote if the accusation is upheld?' Meaghan asks.

Four hands shoot into the air and I smile at Cassidy, Karim, Newton and Smithy. Tomohiro's hand, however, stays in his lap.

'Tomohiro?' I ask.

'I'm abstaining. Sorry, Sair, but without hard evidence I can't vote for punishment,' he says.

'I, too, am abstaining,' Meaghan says. 'Therefore, the vote is won by the defence five to four and the accusation is quashed.'

I can't believe it. I feel like the air has been punched out of my lungs.

'This is so fucked,' Smithy says flatly.

I can still smell Declan's oppressive body odour and feel his hands on my breasts.

He smiles at me arrogantly. The all-powerful king, comfortable on his throne.

I turn my back on the group and walk out into the forest.

'Sarah, where are you going?' Cassidy calls after me.

I don't answer, just keep walking. My feet take me to the weather tree and only as I begin to climb, do I realise that someone has followed me.

'Do you need a boost?' Newton asks gently.

I nod and he proffers his hands as a stirrup so that I can reach the first branch.

We climb together in silence, only stopping when we're at the very top, looking out over the mountains and forest.

'Looks like there's gonna be another storm,' he says, gesturing to the horizon.

Huge black clouds stretch from east to west and occasional lightning strikes pierce the darkness.

'I'm leaving,' I say suddenly, catching us both off-guard.

'What do you mean?' he asks.

'I can't stay with him. He's vile, he's manipulative,' I say, as the thought solidifies in my head. 'I'm gonna leave, start again.'

'But what about everything we've built?' he probes.

'We can build it again.'

'Even if some of us follow you, there won't be enough of us to reach genetic diversity criterion.'

'We don't have enough people to meet GDC as it is,' I rebut. 'At some point we're gonna have to look for other survivors, or other R.E.Generation units. I'm leaving tomorrow.'

'Okay, if your mind is set,' he says tenderly.

'It is.'

He looks at me kindly. 'Then I'm definitely coming with you. I'll tell the others.'

'Just the Epsilons and Etas,' I say hastily. 'I don't want Declan catching wind of the plan.'

'I get that,' he says hesitantly. 'But the more people we take, the better it will be for us.'

I nod thoughtfully. 'I'll tell Tomohiro then. I reckon that I can trust him. Plus, he's on the inside with the others. I'll trust him to tell anyone else that he thinks might come.'

'Are you sure that's wise?' Newton asks.

'Not really. But Tomohiro is wise. He's far smarter than he lets on. We'll have to trust him.'

We sit in silence for a few moments longer, watching the weather unfurl in front of us.

I turn to Newton to clarify the plan. 'Tell anyone who's coming to be ready to leave by dawn tomorrow. We meet a hundred metres into the forest behind the Shit House.'

He nods.

We climb down the tree and return to the others.

It's hard continuing like normal, but I throw myself into the work to distract myself while I plan more details. We work on the final stretch of the aqueduct but have to stop mid-afternoon as the wind picks up, warning of the approaching storm. As we struggle back to camp, fighting into the wind, we are whipped by thousands of tiny leaves, spinning through the air.

Ayesha stops beside me and plucks a leaf from my hair where it has got stuck. The leaf is long and slender and has a bulbous nodule on one end. Ayesha pops the nodule open, revealing a seed inside.

'More food to try,' she says placing the seed in her pocket. 'This must be how the trees have spread so far. There's no insects or birds to spread the pollen or seeds, so they rely on the strong weather to reproduce.'

'Genius,' I say with a smile.

'We should catch as many of these as possible,' she says. 'Hopefully they're not toxic.'

Once we're back at camp, we hang a large sheet of fabric between two trees, fastening it securely at each corner. Many of the seeds fly into the sheet and collect at the base, but many more fly straight past.

Soon the wind dies down, but after climbing the weather tree again, Newton assures us there is worse storm to come.

I avoid telling Tomohiro about our plan until late in the day, to lower the risk of the information spreading.

Like Cassidy and myself, Tomohiro has decided to stay in the original triangular hut that he shared with the other Alphas. He finds manoeuvring his chair into it easier than the slightly raised, newer buildings. Declan and Nat have moved into the leader's hut and the other Alphas have moved into one of the large huts with Tiffany and Poh.

After dinner, Tomohiro wheels himself towards his hut. I follow him and invite myself inside. He smiles at me as I cross the hut to sit on a pile of pine needles he has collected as a sort of pillow.

'What's going on, Sarah?' he asks with a quizzical smile. Since his initial paralysis, some of the function of his face muscles seems to be slowly returning, and his speech is slightly less slurred.

'I'm leaving,' I tell him, cutting straight to the chase.

'That makes sense,' he says. 'I was wondering who would be the first to go.'

'What do you mean?'

'There's too much tension. It was gonna bubble over eventually. If I wasn't in this chair, I probably would've left already myself.'

'Then come with me,' I offer. 'Newton's coming, and hopefully some of the others.'

He shakes his head sadly. 'I can't.'

'Of course you can,' I implore him.

'Look at me, Sarah. I'd only slow you down. Plus, Tiffany and Jay won't come, I can tell you that. We're finally getting somewhere with my rehab. I have to stay with them.'

I take a deep breath. 'I probably should've told you this a while ago, but Declan and Nat asked me to condone killing you. Eliminating the dead weight from society or some bullshit.'

His eyes furrow as I continue.

'They figured if I said yes everyone else would be on board, but of course I told them where to shove it. If you stay here, I won't be able to protect you. Who knows what Declan will do.'

'That's messed up,' he says shaking his head. 'Thanks for telling me.'

'So you'll come with us?' I ask hopefully.

He shakes his head again. 'I've made up my mind. I'm staying.

Someone needs to keep an eye on Declan. Maybe I can keep him from hurting the others.'

'You're amazing, you know that, right?'

He smiles at me.

'Would anyone else come?' I ask him.

'Maybe, I can ask around.'

'Please do. But keep it subtle. Declan can't find out.'

'Of course,' he says with a conspiratorial wink.

'Tell anyone who wants to come to meet us near the Alpha pods.' I give him misinformation so that if word does make its way back to Declan, he won't find the main leaving party.

'No worries,' he says. 'But one more thing. Take my chair.'

'We can't do that, you need it,' I say taken aback.

'And so do you. We have no large hike packs. You'll be able to carry more if you load the chair. Nat can always make me another one.'

'If you're sure,' I say tentatively, seeing the sense in his proposal.

He smiles. 'I am. Take it with my blessing. But come and get it in the morning. I'll need it tonight.'

I kiss him on the forehead and go to leave the tent, turning back at the entrance curtain.

'Thanks, hero. I'm gonna miss you.'

'You too, Sair,' he says with a bow of his head. 'Take care of the others.'

'Same to you.'

ALIYAH

Jacques folds his arms across his chest and leans back in his chair. 'I need you to drop this story, Aliyah,' he says bluntly.

Jacques, Aliyah's boss and editor, has always been stern but fair. She can see why he thinks the story has no legs, but she needs to convince him there is a bigger picture. More at play here.

'I don't think you understand,' Aliyah says calmly. 'We're not just talking about street orphans disappearing. People are reporting stolen children. Kidnapped from public places and right out of homes.'

'So why is there no evidence? Why aren't the police concerned?' he asks.

One of the lights in the ceiling is flickering and the hum of the air conditioner is a constant irritation. Aliyah glances at the light and pinches the bridge of her nose, trying to concentrate. 'These are professional jobs, Jacques, an organised syndicate. Who knows, maybe the police are involved, or maybe they're being paid to turn a blind eye.'

He swivels in his chair, giving a huge pile of papers beside him his full attention. 'I won't tell you again. Drop this story, Aliyah.'

Aliyah walks out of his office, more determined than ever to crack this story wide open.

She visits the families of the children who disappeared. Every story is the same and yet different.

Some babies vanish from their cots in the middle of the night. There are never any signs of forced entry, which indicates that the kidnappers somehow have inside information. Aliyah tries to find a link between the affected families, but nothing stands out.

Other children have been taken in crowded areas. In bazaars and markets, where there are throngs of people, young children are whisked away before their parents even realise what is happening. There is even less pattern to these abductions, seeming to happen at random. Opportunistic attacks on unsuspecting victims. One family even tells her that they've heard of women disappearing in the same way.

Aliyah is frustrated that she can't decipher the criminals' master plan, but she feels good about the work she is doing. Many of the families are eager to share their stories and report that they have no one else to talk to. They tell her that the police just don't want to know. As soon as anyone mentions the word kidnap, they are turned away.

Aliyah receives a call from a new family, but because Jacques has asked her to drop the case, she can't visit them until after work.

When she arrives at their house, she realises it is the same home she visited a few weeks earlier as part of a different news story. It's so hot outside that she works up a sweat just walking from her car to the front door.

'Oh, it's you again,' the mother says when she opens the door. She looks like she hasn't slept in days. Her dark curly hair is unkempt, and her clothes are creased and crumpled. Streaks of mascara running down her cheeks almost blend with her dark skin, but act as testimony to her recent loss and ongoing grief.

'It's Sybella, isn't it? May I come in?' Aliyah asks.

'Yes, and yes,' Sybella says curtly.

Aliyah enters and shows herself to the lounge. She is relieved that the house has good air-conditioning.

'Take a seat,' Sybella says. 'Tea?'

'I'd love one.'

Aliyah tries to remember what the mother's previous story was about. It was something mundane, like a lost pet. No, it was a stain on the ceiling that looked like the president of Namibia.

'I wasn't going to get in contact with the newspaper, but one of my friends told me you were asking around about missing children,' Sybella says. She carries a tray laden with cups, biscuits and a pot of tea.

'What's your story?' Aliyah asks.

'You and your photographer colleague visited and about a week later, Jozette disappeared in the middle of the night.'

'I'm so sorry to hear that. I hope you don't think we had anything to do with it?' Aliyah asks.

'Well, I don't know. I don't trust anyone anymore.'

'Me neither. That's why I'm trying to get to the bottom of this. If you don't mind me asking, what kind of security system do you have?'

'I have a button key. I know the technology is a bit outdated now, but I can't afford an upgrade. Plus, people kept their cars safe with the same technology for years before we put it on our front doors.'

'Have you checked to make sure that you still have all of your keys?' Aliyah asks.

'Of course.'

They chat for another half hour before Aliyah leaves. She heads back to the office and lets herself in. She is surprised to see her boss is still in his office.

'Just got some work to catch up on,' she calls to him. 'Don't mind me.'

She cross references all the families that have been victims of the kidnapping with the newspaper's archive. She is surprised to find that most of the families who had children abducted from their homes have been featured by the newspaper in recent months. She double checks the dates and a feeling of dread starts to seep through her entire body.

She takes off her shoe and checks under the innersole. Her back-up plan is still secure.

All the house abductions occurred within three weeks of the families being visited by newspaper staff. She makes a list of the reporters and photographers in attendance at each visit. Everyone appears to have been present at one point or another, but there are three names that keep coming up. Lethabo, Tshegofatso and Mpho, the photographer who visited Sybella's house with Aliyah.

She collates the information onto a single sheet of paper before sending it to the printer. She collects it with shaky hands and walks to Jacques' office.

'Jacques, I found proof,' she says laying the paper in front of him. 'Employees in this office are scouting houses, so that kidnappers can return and snatch their children without alerting the parents.'

Jacques stares at her silently, so she keeps talking.

'I think all of the houses must have old security and the kidnappers have worked out how to unlock the electric doors with a master key.'

Still no response from Jacques.

'Did you know?'

'I told you to drop this case,' Jacques says coolly. 'There's no money in print media anymore. There hasn't been money in print media for decades.'

'But, but,' Aliyah stammers. 'Kidnapping children. Really?'

'Well, actually, we've been looking at expanding our operations recently. Have you had your fertility checked of late?' he asks menacingly.

She shakes her head.

'We'll have to see to that then.'

She starts to back away, but he moves incredibly quickly. She turns to run but he pounces on her and pins her to the ground.

He hisses into her ear. 'People in America will pay handsomely for a fertile African bride. I hope for your sake that you can have children, otherwise I won't have a use for you anymore.'

Jacques blindfolds Aliyah and bundles her into the back of a vehicle. The heat of the day hasn't dissipated, and Aliyah swelters in the small, enclosed space. They drive for at least an hour, before Aliyah is dragged out of the vehicle and across a cold concrete floor. She is thrown down and she hears a heavy door close behind her. Hands grope at her and remove the blindfold from her face. She is in a dark room, but she can make out the figures of at least seven others. It is warm and sticky in their room, which doesn't appear to be connected to the air conditioning.

She is rolled over and the others untie the binds holding her hands and legs together.

'Thank you,' she mutters.

'What's your name?' someone asks.

'Aliyah. Yours?'

'Hannaleen. This is Itha and the rest are children,' Hannaleen says.

'How did you end up here?' Aliyah asks. She takes off her shoe and removes the object hidden inside.

'Most of us were abducted from markets. There was another lady, but she didn't come back from her fertility test.'

'Are there any babies?' Aliyah asks, soaking up as much information as possible.

'Yes. But they keep them in a separate room. They need more care and can't run away,' Hannaleen explains.

'And what do they do with you? Well, I mean us.'

'Once there are enough of us, they will ship us across to America where rich families will 'adopt' the children and rich men will take us as their wives.'

'It's barbaric,' Aliyah says, disgusted.

At that moment, the heavy door starts to swing open and Aliyah quickly stashes her device in her bra.

'Aliyah, come with me,' Jacques says.

'Good luck,' Hannaleen calls.

Aliyah stands up and walks to Jacques. He closes the door behind her and starts to walk away. Aliyah breathes in the cool, conditioned air as she follows Jacques obediently.

'You're a hideous man, Jacques,' she says accusingly.

'No, I am just a man,' he replies. 'We all do what we need to do to survive. Profit is profit is profit.'

'How big is this place?' Aliyah asks.

They walk through grubby corridors, passing guards and countless rooms with who knows what inside.

'About four hectares. We've got more than two-hundred million dollars worth of stock, ready to be shipped.'

Aliyah smiles – people love to brag when they think they're in control.

She stops walking and waits for Jacques to turn around. She pulls the device out of her bra and holds it behind her back. Jacques finally notices that she isn't following him and turns to face her, looking bored.

'Come–' he starts to say, but she cuts him off.

'You didn't really think that I would investigate a complex criminal gang without some sort of security, did you?'

He looks at her, perplexed.

She shows him the device and snaps it in half. A small orange puff diffuses into the air and she smiles.

'What the fuck is that supposed to be?' he asks.

'Remember that bio weapon story you had me working on last month?' she says with a slight smile. 'Well, this is it. A living organism that seeks out and attaches itself to mature Y-chromosomes. Once it's found what it is looking for, it will multiply rapidly and start to attack the surrounding cells, devouring human flesh until there is nothing left to eat.'

Jacques twitches as some of the micro-organisms start to attack his peripheral nerves.

'The packet says that the range is about one hundred metres in all directions from the point of release, but I imagine with the ventilation system that this place has, every man you have working for you will be dead in the next ten minutes.'

Jacques pulls a gun out from a hidden holster and points it at Aliyah, struggling to keep it level.

'You're dead, bitch,' he says.

She smiles at him. 'I know. But so are you.' She shrugs.

He pulls the trigger and Aliyah collapses to the ground. As blood pours out of her, the pain in her chest is numbed watching Jacques writhe in agony, and it isn't long before they both lie still and lifeless.

SARAH

CHAPTER 10

I leave Tomohiro's hut feeling sombre and walk the short distance to my hut before ducking inside. Cassidy is changing the heated stones, making sure that the eggs stay warm. Ultimately, she's been just as diligent as I have about caring for them.

'Cass, I need to tell you something,' I say, sitting down opposite her.

She finishes rearranging the blanket so that there are no drafts getting to the eggs and looks up at me quizzically. 'Yes? What is it?'

'We're leaving. We have to get away from Declan before he attacks anyone else.'

'But where would we go?' she asks.

'Anywhere. It doesn't matter. As long as it's not here. We can't make him leave, so we have to leave instead.'

'I can't believe he tried to rape you,' she says looking at me with sympathy.

'I can. And I don't think I was the first.'

'Ayesha?' she says, raising an eyebrow.

I nod. 'And I won't be the last.'

'What's the plan then?' Cassidy asks, her moment of hesitation having passed.

I fill her in on the details and we spend the evening packing our rucksacks with important provisions. The bags are woefully small and by the time we've packed dried food, a few handfuls of the pine seeds, water and medical supplies, there's hardly room for anything else.

'What about the eggs?' Cassidy asks as she replaces the warm stones once more.

'I guess we'll have to leave them,' I say sadly.

She looks at me with angst, but then nods her head in agreement.

Later in the evening, the storm hits with full force. The rain and hail make a thunderous sound as they crash against the walls of our hut. As if in response to the rain, the eggs start to shift amongst the stones and blanket. A faint *tap tap* sound emanates from the nest and Cassidy and I hold each other in excitement as we wait for the eggs to hatch. The progress is painfully slow and after about an hour, only a small crack is visible on one of the shells.

The rain has died down a bit and I decide to make a dash for the toilet. Stepping inside the Shit House fills me with dread. I try not to think about what Declan did, not wanting to give him that power, but the memories materialise on their own.

When I get back, Cassidy looks at me with her typical guilty expression. 'I'm sorry,' she says with a small smile.

'What?' I ask, crossing over to her.

She lifts her hands and reveals a small lizard nestled in her lap. 'It hatched so quickly, and then it just came to me,' she says gently.

I sit next to her and smile at the tiny creature.

Its skin is a mixture of brown and orange scales. I imagine it would blend in well with the pine strewn forest floor.

It sits calmly in Cassidy's lap and she strokes it lovingly. It responds to her touch with a quiet purring sound, eerily similar to a cat.

I reach my hand towards it to pat its scaly skin, but it snaps at my fingers aggressively and I have to withdraw them quickly.

'Sorry,' Cassidy says again.

'No worries,' I say eyeing the lizard cautiously. It watches me too, staring deep into my soul.

One of the other eggs now has a crack along its surface and soon after, a loud cracking sound precedes the egg splitting into two neat halves.

I look at the hatchling, who peers back at me. It appraises me cautiously and I feel like it's judging my worth. Then, without any warning, it leaps out of its shell and lands squarely in my lap. It walks a few circles before finding a comfortable place to settle.

'They're a bit bizarre,' I say softly, not wanting to startle the tiny creatures.

'But super cute,' Cassidy replies. 'What are you gonna call that one?'

I think for a moment. 'Aravis, I think. In honour of Karim's cat, Aslan,' I say with a grin.

'Who's Aravis?' she asks.

'From *The Horse and His Boy*,' I say simply.

She looks at my quizzically.

'She's basically the coolest girl in all of the Narnia stories,' I tell her.

'Oh, okay. I think that's a cute idea, anyway. I might call this one Tumnus then. He's a Narnia character, right?'

'Yes, he is,' I reply with a smile. 'I think that suits him perfectly.'

We wait another hour, but the last egg doesn't show any signs of hatching. While we wait, Cassidy fashions each of us a pouch that we can wear over our necks to carry the lizard babies. After a while, they start to explore the hut and watch each other cautiously. The end of Tumnus's tail hasn't quite formed properly, leaving a short stump, but it doesn't seem to bother him. After Aravis has finished exploring, she returns to me and snuggles into my lap. Tumnus likewise seems smitten with Cassidy.

'They seem to have imprinted pretty hard,' she says.

'Yup.'

Aravis licks my finger and closes her eyes, looking serene.

We try to get some sleep, but the noise of the hail prevents me from settling properly. Furthermore, I'm terrified of squashing Aravis, who has snuggled in close to my belly.

I'm also overflowing with excitement and anxiety about striking out from camp in the morning. My mind races and sleep eludes me for hours.

It feels like I've only just fallen asleep when Cassidy shakes my shoulder and tells me that it's starting to get light outside. How her body knew to wake up, I'll never know. I probably would've slept another six hours if she hadn't woken me.

'I need to talk to Newton. Can you go load Tomohiro's wheelchair?' I ask her.

'Of course.'

Luckily for us, Tomohiro's hut has also been used as the main storage building. Cassidy should be able to find some tools, spare water and food in there.

We leave Aravis and Tumnus snuggled together close to the fire. I peek my head out of the hut and scan the camp. The dawn light bathes everything in an orange hue, but the coast is clear.

I dart across to the hut where the other Etas and Epsilons are sleeping. None of them have woken up yet.

I shake Newton gently by the shoulder and his eyes open slowly.

'Time to go,' I whisper.

He nods and stretches to force his body to wake up.

'Who else is coming?' I ask quietly.

'Everyone,' he says with a smile.

'Everyone?' I ask incredulously.

'Well, everyone in here, anyway.'

I smile. With all the Etas and Epsilons working together, we might just stand a chance on our own.

Some of the others have started to stir at the sound of our conversation.

'Newton, I need you to run to the weather tree and check that the storm has passed,' I tell him.

He nods his understanding and after one more stretch, he rolls onto his knees and stands up. Karim is already moving about, rolling up his blanket and collecting his things.

'Karim, when you're done, can you go help Cassidy at Tomohiro's hut?' I ask him.

'What's she doing there?' he asks.

'She'll fill you in. Everyone else, get everything that you're taking and meet me where we planned. Wait for an hour, but if I'm not there by then, head off without me.'

I sneak back to my hut, but there's still no sign of anyone else up and about.

I grab my pack, sling the pouch over my neck and carefully put Aravis inside. She nestles down into the depths of the pouch and seems content. Thank the gods for that.

I make my way carefully back across camp to the Alpha pods. If anyone else is coming with us, this is where they'll meet. But it's also where Declan will confront me if word has made it back to him.

I place my pack next to one of the pods and climb inside. I pull the lid most of the way closed, lying in wait so that no one can surprise me.

Aravis rests on my stomach and starts making a quiet *ek-ek-ek* noise.

I offer her my finger to try and quiet her down and she suckles on it gently.

I prop myself up on one arm and peer through the gap in the pod. I watch as two figures ghost from the Eta and Epsilon hut into the forest on the far side of camp. Cassidy and Karim emerge from Tomohiro's hut and hurry into the forest, pulling the wheelchair laden with supplies. Cassidy re-emerges and ducks back into our hut to grab her pack and Tumnus.

Seconds later, a female figure emerges from the Alpha and Gamma hut. She looks around the camp nervously and then slinks into the trees. A few minutes later, I hear footsteps nearby and then a pair of hips walk into my view, standing directly next to the pod that I'm hiding in.

She shifts her weight from foot to foot and I can tell that she is anxious and jumpy.

I haven't seen anyone else emerge from the hut, so I slowly open the pod to see Poh standing in front of me, looking back into the clearing.

'Poh,' I say quietly, and she whips around in shock.

'Jesus Christ. You scared the crap out of me,' she says, managing to keep her voice hushed despite her shock.

'Sorry,' I say standing up next to her. 'So, Tomohiro told you then?'

She nods. 'I need to come with you.'

The way she says it makes me need to know more. She keeps glancing back at the camp nervously.

'Why?' I ask simply.

She looks at the ground in despair. 'He raped me, Sarah. Not attempted to. He raped me. He's so strong.'

My instinct is to pull her into an embrace. She holds me close and we stay together for a minute without talking.

'I'm so sorry,' I say 'I'm just… so sorry.'

'It's not your fault,' Poh mutters.

We stand in silence for a further minute. There are no words that can make this better.

'This sounds so harsh, but why didn't you say something yesterday?' I ask her, disengaging from the hug. 'In fact, we could go back now and revote. With you we have the numbers.'

'I can't,' she says shaking her head. 'He promised to make my life miserable if I disobeyed him. I couldn't stand up to him in person. I'm not as strong as you. Please, take me with you.' She seems to be working herself up into a state of terror.

'We will, it's okay. He can't hurt you where we're going.'

'Unless he already has. I just hope he didn't make me pregnant. I don't think he finished, but I'm so scared. I don't think he even enjoyed it, he just did it to prove to me that he could.'

I can still feel his hands on me too.

'Son of a bitch. But let's stop talking about him. It's time to move on.' I notice that she doesn't have a backpack with her. 'Do you have any things?'

She holds up a blanket in her left hand. 'Only this. I couldn't exactly pack a bag with all the other Alphas around. It was hard enough getting out this morning without waking them.'

'Fair enough. Is anyone else coming?' I ask.

She shakes her head. 'Just me.'

'Then let's go,' I say.

'But where are the others?' she asks looking around. 'Surely some of the others are coming with us?'

'They're waiting somewhere else,' I tell her.

'Oh,' she says. I see the hurt in her eyes, realising that we didn't trust her.

'Don't take it personally. Now come on.'

We take a large semi-circular path around the camp, staying within the tree line.

We meet up with the others and are about to set off when I realise someone is missing.

'Where's Meaghan?' I ask, and everyone looks around perplexed.

'I haven't seen her this morning,' Smithy says. 'She sleeps in the changing room.'

'Was she planning on coming?' I ask Newton.

'Yeah. At least she said that she was.'

A terrible feeling begins to seep into the pit of my stomach. What if she's gone to tell Declan? She knows that this is the actual meeting place.

'You guys go, I'll see if I can find her,' I tell the group.

'You can't do that,' Newton says.

'Of course I can. This is my plan and I'm not risking any of you getting fucked up by Declan. But you can't wait here either. Walk due south for a few kilometres and I'll meet up with you later. If I don't show up, just keep on going.'

I look at the circle of faces, some concerned, some excited.

I raise my eyebrows and smile, hoping to look confident, and turn back towards camp without another word.

When I get to the tree line, I see that Ayesha and Jay have risen and are sitting by a small fire they've started in the central fireplace. They have their backs to me, and I make my way carefully towards the Eta and Epsilon hut.

Adrenaline courses through my entire body. I'm wired, expecting to hear a shout of discovery or see Declan marching around the corner at any moment.

I duck into the hut and peer into the changing room. Having expected the worst, I am surprised to see Meaghan sleeping calmly under her blanket.

'Meaghan,' I say quietly, but she doesn't stir.

I shake her by the leg and she still doesn't wake up.

I squeeze up beside her and pat her gently on the cheek.

'Wah wah wah what is it?' she calls, waking with a start.

'Shh,' I say. 'If you're planning on leaving camp with us, you have to come now.'

'What time is it?' she asks, wiping sleep from her eyes.

'Past the time that we should be going. Everyone else is on their way. Didn't you hear them getting up?'

She shakes her head. 'Sorry, I sleep like a log.'

'I can see that.'

She already has a backpack prepared and I help her to throw a few extra things into her blanket which she ties with a big knot.

Before we leave the hut, we check on the state of the camp. Nat and Declan are walking towards the fire holding hands and they settle down facing towards us.

'Shit,' I say. 'Shit, shit, shit.'

'Out the window,' Meaghan says gesturing to a large opening at the back of the hut.

She helps me out first, passes me her things and then vaults out of the window with ease. Keeping the hut between us and the fire, we back slowly towards the trees and only relax once we're a few metres past the first trunks and engulfed by the gloom.

'Nice one,' I tell her with a smile.

She nods in acknowledgement.

We have to skirt around the camp again to get to the southern edge but staying well into the trees, we're safe from being spotted.

We find the place where the others set off from and begin to follow their tracks.

'These are pretty obvious,' Meaghan says pointing to the disturbed pine needles and occasional footprints that mark the passage of the others. There are also deep grooves where Tomohiro's wheelchair has been dragged, laden with goods.

I look ahead into the forest and see that the tracks continue into the gloom.

Meaghan and I find a large fallen tree and drag it across the path, obscuring the obvious starting point of the group. We then spend some time gently displacing some pine needles into the tracks of the wheelchair and the obvious footprints.

Suddenly, we hear an angry shout from the direction of the camp.

'What!' Declan cries in a rage. 'Where the fuck have they all gone?'

I look behind us, checking our work

'That'll have to do,' I say to Meaghan and we set off after the others.

Hopefully, if Declan and his cronies come looking for us, it will take them a while to find our tracks. Even so, I want to put as much distance between us and them as quickly as possible.

Meaghan and I move twenty or so metres off the path made by the others and jog parallel to it, heading south. I have to hold Aravis close to me with one hand to stop her bouncing about too much, but she doesn't seem to mind and stays in the depths of the pouch. My pack bounces around on my back annoyingly. Meaghan is a much better runner than me and keeps a relentless pace. I'm soon out of breath and have to ask her to slow down.

It doesn't take long to catch up to the others who are travelling a lot slower than us.

'He knows that we've gone,' I pant.

'And this thing is leaving a trail that will lead them straight to us,' Meaghan says pointing at the wheelchair. She hardly looks like she's broken a sweat.

The others look back the way they came and see two deep grooves in the soft ground.

'Shit,' Karim says. 'We'll have to carry it for a bit.'

'Where are we going, anyway?' Poh asks.

Everyone looks to me and I feel deeply foolish.

'I hadn't planned that far ahead,' I say sheepishly.

Their faces show mixed emotions. Disappointment, horror, fear.

'We go to our sister unit,' Smithy says firmly.

'And how do we find them?' Newton asks.

'I know the way,' Smithy says confidently. 'Or at least I did. I'm hoping that the terrain hasn't changed too much.'

'We all memorized the location, Smithy, but it'll be like trying to find a needle in a haystack. They're ninety k's away,' I say dejectedly.

'Ninety kilometres as the crow flies or one hundred and eighty my way,' he says with a smile.

'How is going twice as far better?' Newton asks.

'And why do you have *your way* to get there?' Meaghan adds sceptically.

Smithy blushes, his pale cheeks turning rosy in an instant. 'My girlfriend is there. I made sure I memorised a route,' he says. 'We follow the landmarks and we'll be able to find them.'

'If the landmarks haven't changed,' Meaghan says crossing her arms across her chest.

'Well, does anyone else have a better idea?' Smithy says, throwing his hand in the air in exasperation.

Everyone shakes their heads and we stand in silence for a minute trying to think.

'Smithy's right,' I say eventually. 'We need to find more people and our best chance is the other R.E.Generation unit. We have to take the chance.'

Newton smiles and Cassidy nods tentatively.

'Sounds good to me,' Poh says.

'Meaghan?' I ask, needing her on board.

'Yeah, fine,' she says unconvincingly.

'Well then, Smithy, which way?'

'We head south-west, which will take us down the mountain and into the valley. We then swing north and follow the valley.'

'That's good. We'll be harder to follow if we aren't heading in a straight line,' I say.

'If they're even following us,' Cassidy adds.

Smithy continues outlining the plan. 'After the valley opens up, we keep heading north-east until we reach the big lake.'

'As easy as that,' Meaghan says sarcastically.

'That'll get us about halfway there,' Smithy says with a smile.

Meaghan rolls her eyes and shuffles her backpack into a more comfortable position. 'We'd better keep moving. Let's go.' She pulls out a compass and leads the way.

Karim, Newton, Smithy and I pick up the wheelchair and carry it behind Meaghan as she makes her way downhill.

Cassidy and Poh pick up supplies that fall from the chair and follow behind.

I check over my shoulder regularly and notice a few of the others doing the same, but I don't see or hear any signs of being followed. We walk largely downhill for an hour or so before the ground in front of us starts climbing up consistently.

'Is this the valley?' Meaghan asks Smithy.

'Well, I thought it was more obvious than this. It should've been steeper, and more valley-ey,' he says hesitantly.

'So, you've got us lost already,' Meaghan says crossly.

'No, no. This is right,' Smithy says, trying to sound confident.

'It's probably just changed a bit over the last, you know, fifty-thousand years or however long it's been,' I say, trying to support Smithy. 'Cut him some slack, Meaghan. We're all in this together, remember.'

'And ain't that the shame of it,' she says sardonically.

Her attitude has left much to be desired since we left camp. To be completely honest, I'm starting to wish that I'd left her sleeping. Then I remember that she had a daughter before she was chemopreserved. We all have our baggage.

We swing to the north and begin the trek proper, placing one foot in front of the next. We decide that we don't need to carry the wheelchair anymore and return to pushing and pulling it across the carpet of pine needles.

The air is hot and sticky at this lower altitude and hiking in the humidity is tiring and oppressive. The trees come and go, with no new variations to break the monotony. The only thing that I can use to demonstrate our progress are the myriad of different mushrooms that pepper our path. Many of them look old, collapsed or eaten by grubs, but a few new mushrooms are already poking their caps through the soil after the night's storm.

I find myself plodding along next to Smithy. He strides confidently and regularly scans the forest on either side of us. I realise that I've become a bit too comfortable in the forest and have been negligent about looking out for dangers.

'I remember you talking about your girlfriend in training,' I say to Smithy. 'I assumed that she was in one of the missing Gamma or Epsilon pods.'

'Nah, she's a Theta,' he responds with a rueful smile. 'All through training, we assumed that we were gonna be together, her being a Theta and me an Eta. But then at the last minute, they told us they were dividing by alternate group. Alpha, Gamma and Epsilon with Eta, and then Beta, Delta, Zeta and Theta together. It almost broke us. She was this close to pulling out of the whole program.' He holds his fingers millimetres apart to emphasise the point.

'What made her stay?' I ask.

'I promised her that whatever happened I would come find her,' he says softly.

'That's so sweet. I guess needing to leave camp worked out okay for you then.'

He nods and grins cheekily.

We walk for a few more hours before stopping for an early lunch. Many of us skipped breakfast and are keen to dive into our food stores, but with only the limited amount that we can carry we have to ration ourselves.

Karim drops his bag to the ground with a metallic clunk. 'Who's got the food, then?' he asks.

'I've got some,' Cassidy says, rummaging around in her backpack.

'What've you got in your bag if you haven't got any food?' I ask Karim.

'Nothing much. Just some in-case-of-emergency kinda stuff.'

I'm surprised that he doesn't give me a straight answer and am suddenly suspicious of what he might've packed. However, he's done enough to earn my trust, so I decide to let it go.

I take out a handful of pine seeds still attached to their dispersal tails and start cracking them open. Aravis, who has slept most of the morning, stirs and pokes her head out of the pouch to see what I'm doing.

I offer her a seed carefully and she takes it from between my fingers and swallows it gleefully.

'What's that?' Poh says, pointing to Aravis.

The others all turn to look as well.

'Um, this is Aravis,' I say hesitantly. 'She's a pet.'

'She's a lizard,' Smithy says shocked.

'She's food,' Meaghan says flatly.

As if aware of all the attention, Aravis disappears back inside her pouch.

'She is not, and *will not* be food,' I say sternly. 'We hatched her, and we intend to raise her. She can be useful.'

'We?' Meaghan asks incredulously.

'Cassidy also has one,' Karim says, pointing to the pouch hanging around her neck.

Cassidy subconsciously raises a hand to hold Tumnus within the pouch.

'As if trying to feed seven humans in the wild wasn't enough, you want us to feed two pets as well,' Meaghan says angrily. 'This is ridiculous.'

'No, it's not,' Cassidy says protectively.

'It kinda is,' Smithy says, shrugging his shoulders apologetically.

'Once they're a bit bigger they'll be able to fend for themselves,' I say, trying to make them see sense. 'They're natives to this land, we're not. They know how to survive. We can learn from them.'

Karim and Newton nod along as I speak. I thank the stars that they both woke up from their chemo-pods because they understand how I think. Having them and Cassidy always on my side fills me with confidence.

'Plus, you saw how big they can grow,' I continue. 'Once they've grown, they can pull this bloody wheelchair for us. Plow the earth. Maybe even give us more eggs.'

'And what makes you think you can tame them?' Poh asks.

'They're not dangerous. Newton says they're herbivores,' Cassidy says. 'And they're super sweet.'

'A compelling argument,' Meaghan says sarcastically.

'Can I see one?' Newton asks. His face shows all the excitement of a scientist on the brink of a breakthrough.

'You can try,' I say. 'They're a bit protective.'

I carefully reach inside the pouch and lift Aravis out, holding her on the palms of my hands. She already fills them, and her tail hangs over the edge. Some orange patches on her back catch the light as she moves.

'Incredible,' Newton says as he approaches slowly.

Aravis turns around in my hands before she spots the pile of seeds beside me and clacks her teeth towards them twice.

I transfer her to my lap and pick up a seed. I hand it to her and she eats the whole thing greedily, leafy tail and all.

'Hello, little darling,' Newton says sitting down beside me.

Aravis watches him carefully, her tail flicking occasionally.

'What did you say her name was again?' he asks.

'Aravis,' Karim says from my other side. 'It's a C.S. Lewis name.'

I smile at him. 'Like Aslan.'

'Like Aslan,' he repeats. 'Thanks, Sarah.'

Newton picks up a seed and offers it to Aravis, but instead of accepting it graciously, she bites him on the fleshy part of his hand.

He retracts it quickly, suppressing a yell and shaking his hand in pain.

'Not dangerous, huh?' Meaghan says. She's been watching the whole interaction while leaning against a tree. Smithy and Poh are sitting near Cassidy, who has pulled Tumnus out and is stroking his neck tenderly. They both look intrigued about the small lizard but are keeping their distance.

'No, no. It was my fault,' Newton says in response to Meaghan's question. 'Sair said that she was protective, but I got a bit too excited. Look, it didn't even draw blood.' He proffers his hand to Meaghan who glances at it briefly.

'Whatever, you guys do what you want. But if they turn violent, I'll have no hesitation in killing them both,' Meaghan says coolly.

I scowl at her.

She ignores me and pulls a piece of mushroom out of her backpack. She chews on it aggressively and eyes Tumnus in Cassidy's lap.

I'll need to keep a very close eye on her.

OLIVIA

Olivia feels like banging her head into the table. They've been over this a million times.

'Well, I'm not sure what I'm supposed to do,' the Prime Minister states. 'If we close down the coal-fired power stations, we lose jobs for thousands of people and more importantly, the backing of some of our biggest benefactors. If we leave them open, then the population says we aren't doing anything about climate change. We're going to meet our targets – what more do they want?'

'Gary, you don't get it, do you? They want a government who sets the agenda. Not one that just meets targets we agreed to ten years ago,' Olivia says. 'You appointed me Minister for Climate Change and Energy because of my background and knowledge. I accepted the ministership because you promised that you would do something big.'

'And I meant it at the time, but things have changed,' Gary says.

'Yes, I know. The effects of climate change are being felt earlier and more severely than we ever imagined. That's why it's time to act now.'

'I mean things have changed politically, Olivia. It's an election year. It's not the right time to push through a radical green agenda.'

'It never will be, boss. Close the power stations and close the coal mines. Give them a year's warning. We'll have plenty more renewables up and running by then. Look what they've achieved in Germany.'

'You can't seriously mean to close down the mines, too,' the Prime Minister says impertinently. 'The country would be broke in a week.'

'There's no point closing down our power stations if we just sell the coal to China and India. It needs to stay in the ground. I expect an announcement in the next two weeks, or I am stepping down as minister.'

Olivia walks out of the office, her rage barely contained and burning like a rocket in her belly. What's the point in getting into politics if you can't make meaningful change?

The following night she holds a media conference, outlining her bold plans, but not going so far as mentioning the ultimatum she gave to the Prime Minister. She ignores his phone calls that night, preferring to turn to social media and bathe in the accolades sent her way.

When she arrives at the party meeting the following day, she is completely blindsided by an unheralded cabinet reshuffle.

'The position of Minister for Climate Change and Energy has been handed to Gavin Dudd,' the Prime Minister announces.

Olivia shakes her head and sighs. What a prick. She knows that Gavin previously worked for a big mining company and still has lots of connections in the industry. There goes Australia's chance to be a world leader in the climate change arena.

'But we haven't forgotten about Miss Olivia Fitzgerald, our bright young star,' the Prime Minister says, catching Olivia off-guard. 'Violet, our much-respected Foreign Minister, handed me her resignation yesterday due to her poor health. That is what prompted this reshuffle and I am happy to announce that Olivia will be assuming Violet's role as Foreign Minister. Congratulations, Olivia.'

Someone slaps Olivia on the back and the room is filled with polite applause. Olivia doesn't know anything about foreign policy. What is he playing at?

Olivia stands outside the Prime Minister's office, her hands sweating. A million thoughts race through her head but she can't think of one logical thing to tell Gary. His aide calls her inside and then leaves so that they have complete privacy.

'Olivia,' Gary says kindly. 'How are you feeling?'

'Baffled, to be completely honest,' she says, her brow furrowed.

'Then let me explain. As I said, it's not the right time for us as a party to do what you want us to, but after the election, we will reconsider. Before then, there is a summit in Russia that I am really keen for you to attend.'

'COP36,' Olivia adds.

'Yes. With your background and enthusiasm, I want you to go and fight for a more rigorous agreement. Greater sanctions for countries that don't meet their promises. More ambitious targets.'

'But if we are not being ambitious as a country, how can I demand that from other countries?'

'Tell them that if the world is in it together, Australia will be in the thick of it. But we can't do it on our own.' The Prime Minister smiles and shakes her hand.

She wishes she had his optimism.

A few months later, at the UN Climate Change Conference, Olivia feels totally out of her depth. There are bureaucrats and officials everywhere, but her own aides set up meetings for her and keep everything running smoothly. Her counterparts from the Scandinavian countries like what she has to say and vulnerable countries like the Pacific nations urge her to keep fighting. They have been asking for similar things for years, and now hope that a bigger country such as Australia will be able to get better results.

Her most frustrating conversations are with officials from China, India and Russia.

'We are still a developing country,' the Indian foreign minister states. 'How is it fair for advanced countries like yours to say that we can't use coal to catch up. You did it for years without repercussions and now you want to clamp down on us and slow our growth.'

'That's not it at all,' Olivia says. 'We just know the consequences now. Invest in solar, wind, tidal energy. I don't care, we just have to stop polluting our atmosphere, right now.'

The following day, the grand debate is staged, with every country represented in the one room.

The secretary-general of the United Nations stands up and addresses the crowd.

'I am sure you are all aware of the bold changes that have been proposed to the climate agreement. Rumours have been circulating like wildfire, so why don't we straighten out the proposal here and now. You should each have in front of you documents outlining the proposed changes, and Australia's Olivia Fitzgerald will run through it now.'

Olivia is shaking like a leaf as she stands to address the assembly. Her voice quavers and she hates herself for showing nervousness, but as she continues, she settles into the occasion and speaks clearly and concisely. She proposes amendments to the climate agreement, including prohibiting the construction of new power stations reliant on fossil fuels and rejecting all future licenses to drill for oil and gas.

There are angry shouts in the room as she discusses the plans, but the secretary-general quiets them down and encourages Olivia to continue.

She proposes doubling emission reduction targets and halving the timeframe in which nations are expected to meet these goals.

'The time to act is not now,' Olivia says, winding up for the end of her speech. 'The time to act was ten years ago, and we missed it. But as my mum always said, better late than never. Let's get this done.'

There is rapturous applause from sections of the room, but also many stony faces staring her down. The Minister of Foreign Affairs from China stands up to address Olivia. He looks like an arrogant businessman who likes the sound of his own voice.

'I admire your enthusiasm and attitude, but when did Australia start letting teenagers rant and rave on its behalf. Talk about hypocritical. Minister Dudd just granted a licence to a company to drill for oil off the coast of Western Australia. How do you expect us to listen to you if your own government won't?'

'I don't think your information is accurate,' Olivia replies.

'I have here a media release from the office of the Australian Minister for Climate Change and Energy. Would you like to see it?' he asks.

Olivia's assistant passes her a phone showing the media release that he speaks of, confirming the truth of his words. It is time stamped three minutes ago.

'I have to apologise,' she says. 'I have just been shown the media release and I really don't know what to say about it.'

The remainder of the summit continues exactly as one would expect. The politicians argue and jostle for position. The bigger countries flaunt their power and make concessions based on what they think they can acquire in return, rather than what would help the health of the planet.

Olivia returns to her hotel room completely dejected and unable to see a way forward where real change is made. Her mind races at a million miles and her hands get sweatier and sweatier. She can't sleep, but sometime in the middle of the night, her mind finally becomes calm and she puts a plan in place. It's not a rational plan, because she's not thinking rationally. The futility of it all is overwhelming. She has tried and failed to get the attention of politicians and the press. No one is paying real attention, and no one is making real changes.

The following day, however, she finally gets to be front page news. The paper reads, *Australia's Foreign Minister found dead. Innovative and radical climate policies die with her.*

SARAH

CHAPTER 11

After lunch, I tuck Aravis back into her pouch and we continue on our way.

Although the trees provide plenty of shade, the sun keeps the temperature rising into the afternoon. The warm air evaporates the night's rainfall off the sodden ground, and the humidity reaches near unbearable levels. Fortunately, the way is flat, so we don't have to deal with hills while we perspire.

'Are you sure this is the right way?' Meaghan asks. 'I thought you said we were supposed to be in a valley.'

'Well, it's flat isn't it, so we must be in the valley,' Smithy replies irritably. 'And it doesn't matter anyway. If we keep heading north, we'll hit the lake. We can't miss it.'

In this heat, no one has the energy to argue with him. We begin to spread out as we trudge along.

'Hey, make sure you can always see someone in front of you,' I call out. 'If not, tell them to slow down.'

I receive a few grunts and calls of acknowledgement.

Cassidy and I keep pace together. We occasionally offer encouragement to each other when one of us starts to flag, but otherwise we trudge in companionable silence.

'So, you and Karim, hey?' she says, waking me from my stupor.

'Whaddaya mean, me and Karim?'

'Well, are you a thing or something?' she asks carefully.

'Karim? No,' I reply, caught off guard.

'But you fancy him. How could you not?'

'No, not really,' I answer. I want to say more, but I'm so tired and surprised by her line of questioning that I can only manage short answers.

'But you named your lizard after his cat. That's sweet. And you two get along so well. And he always hugs you.' She looks at me carefully, hoping I'm not having her on.

'I swear to you, Cassidy, I'm not interested in him,' I tell her, finally finding my words. 'He's more like a brother to me. And besides, I like Newton.'

'You do?' she says, surprised.

'Yeah. I think so. He's so kind. I always assumed you were interested in him,' I say.

'Not really. He's all yours,' she says with a wink. 'So, you wouldn't mind if I got with Karim, then?' she asks.

'Not at all, I'd be happy for you both. But I'm not sure that it's just up to us.'

'Of course, of course. Genetic controls and all that. I attended the same training as you, remember.'

'Plus, who knows if they like us?' I add.

We fall back into silence and continue our hike.

'You're absolutely certain, though?' she says, fifteen minutes later.

'As long as he's happy and it works for the group, Karim is all yours,' I say with a smile.

When we stop for a break mid-afternoon, we decide to call it a day.

'There's no point over-exhausting ourselves,' Smithy says pointedly. 'We've got a long way to walk this week.'

We're all thankful to be stopping, but there's no escaping the oppressive humidity. We have to ration our water, so we can't even quench our thirst properly.

We laze around for the afternoon. We discuss building a fire to cook dinner but decide against it. Firstly, because it will indicate to the others where we are, and secondly, it's already too bloody hot as is.

Luckily, we were able to harvest some fresh mushrooms during the hike, so we have plenty that we can eat raw for tea.

As we sit around eating and chatting, Cassidy steers the conversation towards sex.

'I'm just saying, it's probably a good time to talk about it again,' she says coyly. 'What are we waiting for, hey?'

'To be safe and settled maybe,' Poh says.

'But it'll be nine months before we have any kids, and we've survived this long,' Cassidy adds.

'Okay, how do you suggest we go about it, Cass?' Newton asks.

'Everyone picks one name, in their head, that would be their preferred partner. If two people pick each other then they pair up. If not, we can decide some other way.'

'Before we get started,' Meaghan says hastily, 'I just want to make it clear that I am not fussed in the slightest who I'm paired with, because I'm gay.'

We all look at her in surprise.

'Sorry, what?' Cassidy asks.

'I'm gay, so I'll procreate with anyone,' Meaghan says simply.

'But your daughter?' I ask in shock.

'I think we're all allowed to make mistakes when we're young,' she says. 'Not that my daughter is a mistake. She's wonderful. But sleeping with her father was foolish.'

'I'm sorry, Meaghan, I didn't know,' I say, unaware of what else to say.

'Why are you apologising? I don't need to apologise and neither do you. It's just part of who I am.'

'Yes, of course,' I say feeling foolish.

'Why did you want to join the R.E.Generation Project then?' Karim asks.

'Same reason as the rest of you. Adventure and to help humanity survive. Just because I'm gay doesn't mean I can't carry children.'

'Of course not,' Karim says hastily.

'And there's gonna be gay kids in this world, and I want them to know that it's okay from the outset,' Meaghan says proudly.

'Here, here,' Newton says.

'Good for you,' Poh adds enthusiastically.

Meaghan nods her appreciation and smiles at the group. It's good to see her calm and engaged.

'So has everyone decided on their name?' Cassidy asks, bringing the conversation back to her desired topic.

'Give me a sec,' Poh says. After a moment's pause she nods. 'Go ahead.'

I look across at Newton and smile. He smiles at me before looking away. I've never had as much trouble reading a guy as I do with Newton.

Sometimes I feel like we just click, and that he might feel something for me too. But oftentimes, I feel like he's deliberately distancing himself from me and pushing me away.

'Alright, Smithy, who did you choose?' Cassidy asks.

'My girlfriend. I'll wait for her, thanks,' he says.

Cassidy laughs nervously. 'Of course, sorry. I forgot.'

I didn't realise that she felt so strongly about Karim but seeing the way that she's acting, I can tell she's very nervous about the possible outcomes of this conversation.

'Newton?' I ask, taking control of the conversation and giving Cassidy a chance to catch her breath.

'My choice is Meaghan,' he says looking at her and avoiding my gaze.

'Interesting,' Smithy says slowly.

I feel a dagger punch me in the chest. Fair enough. He didn't have to choose me. He could've chosen Poh or Cassidy. But why would he choose Meaghan when she's just told all of us that she has no interest in men?

'I presume you're okay with that, Meaghan?' Karim asks.

'Yes, of course,' she says plainly.

'Meaghan and Newton, the first pair,' Smithy announces. 'I guess that leaves you, Karim.'

I see Cassidy stiffen up in nervous anticipation.

'Cassidy,' Karim says simply.

'Karim!' Cassidy says, the word bursting from her mouth. 'I also choose Karim.'

'Aw, how sweet,' Meaghan says.

'And so, the worst kept secret is out,' Newton says with a grin.

'Wait, you knew?' Cassidy asks bewildered.

Several people nod, including Karim.

'I also would've chosen Karim,' Poh says. 'If it makes any difference.'

'It doesn't, really,' Smithy says bluntly. 'But what about you, Sarah? Who'd you choose?'

'It doesn't matter,' I mumble, getting up. 'Congratulations to the lucky pairs.'

I walk away from the group and walk short laps, wanting to stay close to the others. The evening has brought with it some cooler air, which is sweet relief.

Karim and Cassidy disappear into the forest holding hands. When they return half an hour later, I can't help feeling a pang of jealousy.

After dinner, Cassidy demonstrates how to weave a basket from thin, green pine branches. We each have a go, with differing levels of success. By the end of the evening, we have one shoddy basket made that everyone contributed to.

'It gets easier with practice,' Cassidy promises.

We sleep in the open, with a watch set to make sure that nothing surprises us in the night. We haven't seen anything to be afraid of, but in a new part of the forest, we can't be too safe.

I'm lucky enough to escape sentry duty and despite the lingering heat, fall asleep quickly on a pile of soft pine needles. I wake in the middle of the night and after resisting the urge to get up, finally give in to my bladder and go for a piss.

I walk a few metres away from the others and squat down. I look up and, through the branches of the trees, marvel at the night sky. The stars shine brighter than they ever did before, and there seem to be many more of them.

'Beautiful, isn't it?' a voice says from the darkness.

My first instinct is to stand up and cover myself, but I'm still mid-stream so I have to stay put. I scour the darkness in front of me and can just make out a figure leaning against a tree.

'You startled me,' I say. 'What are you doing?'

'Keeping watch,' Meaghan says.

There should be more awkwardness about the fact that she's watching me go to the toilet, but her casual indifference helps put me at ease too.

'Of course,' I say.

'Best thing about it is getting to watch the stars and moon,' she says, staring through the canopy above us.

I stand and pull my pants back up, preparing to go back to bed.

'I'm sorry for what I said about your lizard,' she says. 'I was hot, and I don't trust Smithy about this wild goose chase we're on, so I was already grumpy.'

'Thank you for apologising. That's very big of you,' I say sincerely.

'She seems sweet, the lizard.'

'She is,' I say proudly. 'And something tells me we haven't even begun to see what she and Tumnus are capable of.'

She shifts to a more comfortable position as she settles in for a few more hours watch. 'We're gonna need all the help we can get on this crazy journey.'

'You didn't have to come, you know. You could've stayed with the others.'

She scoffs. 'And stayed with that Declan prick without you guys for protection? I don't think so. I never thanked you for coming back for me, did I?'

'It was nothing,' I say.

'Well, thank you anyway. You could've left me with that jerk, but you risked getting caught to wake up this sleepy head. I really appreciate it.' She smiles at me and I realise that I'd been beginning to paint her with the same brush as I had Declan, which is wildly unfair on her.

She is a complex character, with flaws and baggage like the rest of us. However, she's also got an awful lot to offer in this new, unknown world.

'Well, I really appreciate you being a part of this journey,' I tell her. 'Like you said, we're gonna need all the help we can get.'

After her long stint as sentry, Meaghan takes a while to get moving in the morning, but after a dull breakfast of mushroom and pine seeds we head off.

During the morning, we make good pace as the temperature is more manageable and our enthusiasm is high. However, as the afternoon comes around, we begin to drag our feet and slow down as we trudge forever north. Conversation dies down amongst the group as we each retreat into our own bubble and concentrate on putting one foot in front of the other. We take it in turns to push the wheelchair, each handover providing a small moment of tension as nobody wants to deal with the extra burden.

I'm nearly ready to call a halt to the day's marching when the ground in front of us starts to climb steeply. We trudge on for a few hundred metres but ultimately, the hill is the last straw and we decide to set up camp for the night. Everyone except Newton gladly collapses onto the ground.

'I'm gonna find a tree to climb and check the weather,' he says. 'We don't wanna get caught out in another storm.'

'Do you want company?' I ask, mustering the last of my energy stores.

'Sure.'

I almost rescind my offer immediately as he marches up the hill, which becomes steeper and steeper. Only once we're on the ridge at the top of the slope does Newton select the largest tree. He helps me up onto the first branch and while I wait for him to join me Aravis climbs out of her pouch, leaps onto a branch and scuttles up the trunk of the pine.

'Looks like you chose a good tree,' I call to Newton. 'Aravis approves.'

We climb in silence, and I can't help but remember the first time that we climbed a tree together. The jaw-dropping views from the top and the feeling of sharing something special, just the two of us. Did I misread the situation? Did I project my own feelings onto Newton, and imagine he felt something for me, too? After all, he rejected me then, and he's rejected me repeatedly ever since.

'Why Meaghan?' I ask him as we near the top of the tree.

He doesn't ask me to clarify my question, so I assume that he understands my meaning. 'Because she is like me,' he says without breaking his climbing rhythm.

I stop climbing and look at him. 'What do you mean?'

He keeps climbing and then stops a metre or so above me. I scramble to catch up to him and wait for him to elaborate.

'When Meaghan told us that she was gay, I felt a great weight lift off my shoulders. *I'm not the only one*, I thought.'

'Uh huh,' I say starting to piece it all together.

'Since we're both gay, doesn't it make sense that we have children together?' he asks.

Hearing him say those words punches me hard right in the gut, and yet strangely doesn't change the way that I feel about him. But it does mean that I'll never have a chance. That I never *had* a chance.

'Of course, perfect sense,' I say calmly. 'I just hoped there was something between us is all. It's just something I'm gonna have to deal with.' I turn away from him and wipe a small tear from my eye.

'I'm sorry if I led you on. I tried to quash that once I realised how you felt,' he says, looking concerned.

'You're right. You did. I just held onto a hope, a dream, I guess. It's not your fault.'

He looks at me with care in his eyes. 'I'm sorry.'

'You've got nothing to apologise for,' I say. 'I just hope that these feelings I have for you can transform into love for a friend. That we can have fun together, laugh together.'

'I'd like that,' he says.

He moves closer to me and tries to give me a hug, but the branch I'm on makes a threatening, creaking noise and he almost loses his balance as he steps back rapidly.

I shoot an arm out, grabbing his shoulder to stabilise him. We both burst out laughing, the terror of the moment having passed but the adrenaline still pumping through our bodies.

'The first of many,' he says with a smile and continues to climb, checking each branch carefully as he goes.

Newton reaches the top of the tree ahead of me. 'Shit,' he says, looking out over the treetops.

I carefully climb up beside him. Elevated on the ridge we have great views of where we've been, and what lies ahead. What we see is totally unexpected.

I would estimate about thirty k's ahead of us is a bay with a large island, beyond which we can see the ocean. Further around to the right is another, larger bay. Pink, fluffy clouds slowly skid across the blue sky, but nothing else moves from horizon to horizon.

'This was the city,' I say in awe.

'Yup, it's all gone,' Newton says in a hushed voice.

Before we were chemo-preserved, these two bays were surrounded by a bustling city, populated by more than ten million people. Now, there is no sign of them.

The people and infrastructure have been reclaimed by nature. The vast pine forest appears to end a few kilometres from the ocean, replaced by a large plain of shimmering grasses. On the coastline itself, large palms appear to have established themselves as the dominant flora.

Looking behind us, we have a fantastic view of the mountain range. The large mountain that our unit was built upon rises above the rest. On its right-hand side, a large valley separates it from the rest of the range.

'I guess that's Smithy's valley,' I say, pointing back essentially the way that we came.

'I don't understand,' Newton says. 'We thought we were following the compass north, but we've come south instead.'

‘I think I understand,’ I say. ‘Come on, we’d better go tell the others.’

‘Would you like the bad news or the bad news first?’ I ask once we re-join the group.

‘The bad news, I guess,’ Cassidy says.

Despite their tiredness we have everyone's attention, five curious faces eager to know what we’ve found.

‘We've been walking in the wrong direction for two days. We have to turn around and head back that-a-way,’ I declare calmly.

They groan collectively and Meaghan scowls at Smithy.

‘But how is that possible? We've been heading north, right?’ he asks.

‘Wrong,’ Newton says. ‘We've come fifty k's south and are on the outskirts of the city. Or what was the city. That's the other bad news. Everything is definitely, one-hundred percent wiped out.’

‘But how have we been going so wrong?’ Meaghan says. ‘I know how to use a compass.’

‘The poles,’ I say. ‘The magnetic poles. They must've switched. It's happened before and scientists were always predicting it was gonna happen again.’

‘So, we did walk north, according to the compass. Which is actually south in our minds,’ Karim says.

‘Exactly.’

‘Bloody hell,’ Meaghan says grumpily. ‘So we've been hiking for two days for nothing.’

‘And we're gonna have to go past the old camp again.’ Poh adds.

‘Not quite. We can see the valley that we were always supposed to be in. If we head for that we'll slip past the camp a few kilometres to the east. Well, I mean, to the west nowadays.’

'I'm so confused,' Smithy says.

'We'll get used to it,' Karim says with a smile.

MING

The full moon hangs over their heads as Ming and his unit carefully work their way along the roadside. The trees that line the dirt track provide them with cover and open farmland spreads out on either side of the road. Still, you could hide an army in there if you tried hard enough, so they remain forever alert. Scanning the crops for signs of movement. Half-expecting a bullet to come whizzing through the air to take one of them out.

They drip with sweat as the oppressive heat hardly lets up over the course of the night. After an hour or so of marching, they hear a gunfight break out ahead of them. Ming signals to his lieutenant, May, who is in charge on the opposite roadside, and the unit stops as one. Ming listens carefully, appraising the situation. He checks his Comms unit and notes the position of the battle. The Comms unit of the captain involved automatically beamed out her location the second that the first bullet was fired. The firefight is a fair distance away and Ming doubts they will be able to close by morning to help their comrades. He signals to May and they continue forward, following the sounds of exploding gunpowder.

With all the new military technology and computer analytics, Ming didn't expect to play a key role in this war. But with resources the major driver for the conflict, and both sides wanting to minimise losses of food and other commodities, they have reverted to boots-on-the-ground warfare. Their rifles are more powerful and though they have ConeShells to protect them through the day and state of the art communications technology, a bullet through the skull will still kill any man or woman.

The sky starts to brighten as dawn threatens and Ming calls a stop to their march. The unit has four ConeShells, and they huddle in groups as they are deployed. Keegan lays a curved, pole-like structure on the ground and after activation, it unfolds over Ming's head, forming a translucent dome. The rim of the ConeShell automatically digs itself into the ground, creating a complete seal. There is room for eight people inside each Shell, but they have no real comfort and are cramped for space.

They set about making dinner and someone turns on the small air conditioner, lodged in the wall of the shell, to counteract the heat from their cooker and bodies. The sun has now risen outside and checking the ConeShell's weather monitor, Ming can see that the temperature is already soaring. The oppressive night conditions are nothing compared to the heat the day brings. Global warming has reached such extreme levels, that being outside in the daytime results in rapid death due to heat exhaustion. Ming can't help but hate the previous entitled generations for their greed and stupidity. If only they'd acted sooner and hadn't let it get this bad. Then again, before the world started fighting back, there were areas of Canada where you couldn't be outside in winter for risk of death, so maybe this is just the other side of the same coin.

They sleep through the day, Ming using an eye mask to keep out the sun's light. He's never been able to sleep well with any sort of light around, not even a small flashing light on a computer or charger.

They pack up once the sun has set and the temperature outside has decreased to forty degrees. Ming checks his Comms unit. They're still about fifteen k's from the firefight and they settle into a weary march to close the distance as quickly as possible.

They hear sporadic gunfire as they approach, each side probing and prodding, but not committing to a full offensive. Ming has informed the captain up ahead that they will rendezvous with them shortly, and he wonders whether the enemy's hesitation is also due to incoming reinforcements. After a couple of hours guided by the moon's light, they start to see muzzle flashes. The gunfire shines brightly in the darkness of night.

May and Ming disengage from the rest of the unit and creep forward, hoping to find Yu Yan, the captain of the unit engaged in the fighting. Ming sends her a direct Comms message so that they are not shot by their allies as they approach. Yu Yan sends back her location and soon May and Ming arrive at the frontline.

'What are we looking at?' Ming asks.

'Two or three units,' Yu Yan replies. 'There were definitely only two last night, but we believe a third unit has recently arrived. They have increased the intensity of their attacks in the last thirty minutes.'

'And what do we have?' May asks, while she syncs her holo-device with the one held by Yu Yan's lieutenant. It will show the deployment of their soldiers and the predicted locations of the enemy.

'Now you're here, we also have three units. But there are two more units that should arrive before morning, and three more will arrive tomorrow night if needed.'

'So, we will have superior numbers soon,' May says optimistically.

'Assuming they don't have similar reinforcements,' Yu Yan agrees.

'And assuming that they aren't luring our troops together for a mass attack,' Ming says.

They all look at him aghast.

'What?' he says holding his hands out apologetically. 'It's what I would do.'

Ming calls the rest of the unit to the front line and they hold the invaders off as the fighting intensifies. Crops are sawn down by bullets and trampled by military boots. The harvest coming months before it is ready. The very thing they're fighting over, destroyed in the fighting.

A few hours later, two units of reinforcements arrive, just as Yu Yan had declared, and they start to force the invaders backwards, their weight of numbers winning out.

The sunrise, however, brings a complete ceasefire. A shared understanding by both sides that any person left in no-man's-land will be dead by nightfall. Calls for retreat are made on both sides and they regroup, counting their losses and gathering together so that they can deploy their ConeShells.

Ming imagines it is almost like the stories of Christmas day, way back in the First World War. An unofficial ceasefire where the soldiers decided to stop the bloodshed and sing carols for just one day. It's like that, except without the carols, and it must happen every day in this new hell of a world.

Ming and his troops retreat a few hundred metres and deploy their ConeShells. The translucent fabric unfurls over their heads and the outer ring buries itself into the ground. Keegan flicks on the air-con unit, setting it to low.

Some of the soldiers have started to prepare a meal when a bullet suddenly hits the shell. Ming looks around at his comrades, surprise and fear spread across their faces. His mind races, trying to assess the situation and the best course of action. A second bullet hits the Shell.

'Sniper,' someone mutters, fear evident in their voice.

The ConeShell withstands the initial two bullets, but the accuracy of the firing means that the third bullet hits the weakened spot and punctures a hole in the Shell. As they go to retract the ConeShell another puncture appears in the Shell, and then another, and then another.

As the now useless ConeShell is retracted, they all grasp their weapons, ready for the fight. Without protection from the Shell, Ming, and his fellow women and men are as good as dead. But not quite yet. They rush forward keeping low amongst the crops for protection. Two of Ming's men are picked off by the sniper, their accuracy and speed utterly deadly.

Ming can see the enemy task force that haven't yet deployed their own ConeShell and is surprised to see four snipers standing proudly amongst them. There are no other ConeShells nearby, indicating that this is either a rogue crew that ignored the ceasefire, or a suicide mission to take out as many sub-units as possible. Ming has time to glance sideways and can see at least four other damaged ConeShells.

The woman beside Ming goes down, but Ming returns automatic fire, mowing down one of the snipers and two other men. The idiots appear to have had utmost faith in the snipers and haven't even bothered to lift their weapons. Ming and his comrades mow down the sniper task force and, seeing small pieces of paper flying from a dying man's limp hand, he realises they had been betting on how many soldiers the snipers would take out before their sub-unit was overrun.

Enraged, Ming thunders forward, his troops following him into battle. They pass the dead men of the sniper unit and soon find the ConeShells of the enemy encampment. They fire upon them, punching holes and causing irreparable damage. Many people will die this day.

Ming can't help but wonder of the stupidity of it all as the Shells begin to retract, revealing a small, pissed off army. These men only fight because they have starving families at home and they need more land and food to survive. Ming fights to stop these invaders from taking the precious food that keeps his own family from starvation. In the end, someone loses, and everyone is just trying to survive.

By destroying the ConeShells, he has killed these men and women. None will survive the heat of the day. The earth fights back, claiming more victims.

Ming's only hope is that his sacrifice today means that it is his family that survives. His job is done. He keeps firing anyway. A bullet punches his leg. A second goes through his shoulder. Still, he keeps his rifle raised. A moment later, another pierces his skull. Ming topples to the ground, a lopsided smile fixed forever on his worn and tired face.

SARAH

CHAPTER 12

The next two and a half days, we retrace our steps back towards the mountains. While I thought that the route had been flat going the other way, I now realise we'd had a slight downhill assisting us towards the coast. That of course means that our trek is constantly uphill on our way back, and although the gradient is slight, it saps extra energy, needing to climb and climb.

Newton scales trees regularly to make sure that we're staying on course and eventually, we make it into the valley. The trees hide most of the landscape around us, but somehow, I can feel the mountains rising on either side, hemming us in. The most noticeable difference is that it gets darker earlier as the sun sets into a higher, hidden horizon.

On our first full morning in the valley, we crest over a ridge and our route becomes downhill once more. Soon after, we stumble across several small pools, separated by a dry creek bed. The pools are bordered by luscious tree ferns, whose fronds open like umbrellas. They are the most beautiful ferns I've ever seen. Mainly because they're not bloody pine trees or mushrooms.

'You know what this is?' Cassidy asks confidently.

'No, and you do?' I ask perplexed.

'It's the stream from the spring. The original stream. But it's not flowing anymore because we stole the water and redirected it to our camp.'

'You may be right,' I say, nodding my head.

The water is a bit muddy, but none of us hesitate to strip down to our underwear and jump into one of the larger pools. None of us have washed in five days and the collective stink of the group was beginning to get overwhelming.

We consolidate all our remaining fresh water into a few canteens and fill the remainder with water from a pool further upstream.

We find Meaghan hacking into a tree fern with a large knife.

'Looking for critters?' Newton asks.

'Nope,' she says, hacking off a section of pale green flesh from the centre of the tree and holding it aloft. 'Looking for this.' She bites into it and chews the chunk of flesh with vigour. 'Mmm, just like I remember.'

We all crowd around, and Meaghan hacks the inside of the plant apart for us. The tree fern's heart is relatively flavourless, but juicy and refreshing. It has a similar texture to apple but tastes rather like grass.

I go back for seconds and thirds. It's amazing how having a limited diet makes any new food seem like the best thing ever, even if it's plain and watery. Meaghan needs to hack open a second fern to cater for everyone's demands.

We follow the dead stream downhill until it is joined by others and starts to flow again. We tip the muddy water out of our canteens and refill them with clear mountain water.

We follow the stream through the valley for the rest of the day, relishing not needing to climb and being able to drink as much water as we need. We eat fern hearts again for lunch, but by dinner, I'm already over them, and nibble on a selection of mushrooms instead.

The next day, Newton reports that we've passed through the valley and have reached more open country. Smithy instructs Meaghan to lead us due north-east, now actually south-west. We leave the stream that we've been following and hike without any troubles for the rest of the day. We cross lots of small streams and tributaries and late in the afternoon, we are confronted with a large river.

More than fifteen metres wide and flowing rapidly, it represents a significant barrier. Like the smaller streams, the riverbanks are scattered with tree ferns. Smaller ferns sprout from the base of nearby pine trees and grasses grow along the banks where sunlight finally reaches the ground. While the river is an obstacle that we'll need to find a way past, the gap that it provides in the otherwise non-stop tree canopy is a welcome relief. I notice a number of the others smiling at the water as it flows past.

We decide to rest for the night and conquer the river in the morning.

The following day, we split into two groups and head east and west along the riverbank, looking for somewhere to cross. My group and I walk for about an hour upstream before turning around empty handed and hoping that the others have had better luck than us.

Unfortunately, they didn't find anything useful either, except for a section where the river narrows a few hundred metres downstream. We all check it out, but as the river narrows, it also quickens and becomes more turbulent.

'I know that we said that we weren't gonna swim it, but I think it's worth a try,' Meaghan suggests.

'Definitely not here,' Poh says pointing out the rapids.

'Of course, I'd have to find a wider, slower part, but if I can get across with a rope then the rest of you could follow.'

As an ex-Olympic athlete, Meaghan definitely stands the best chance of getting across, but the risk of something major going wrong seems alarmingly high.

'Do you do swimming in the pentathlon?' Karim asks.

'Modern pentathlon,' Meaghan corrects. 'And yes. Two-hundred metres free. I can do this.'

'Okay, but we'll tie the rope around your waist first, we don't want you to be swept downstream,' I say.

As it turns out, we only managed to pack one piece of rope between all of us, and to call it rope is a bit of a stretch. It's a thin cord, about thirty metres long and skinnier than my pinkie. It was designed to last, however, and should easily be strong enough.

Meaghan strips to her underwear and we tie the rope securely above her waist. I take a moment to admire her toned body. She has muscles where I didn't even know muscles could exist.

'You done staring?' Meaghan asks me with a wink.

'Sorry, I… you look great,' I mumble.

'Thanks,' she says with a smile. 'Let's do this, shall we?'

We feed the cord around a tree and the rest of us all take up a handful and brace ourselves. Meaghan walks tentatively into the water, occasionally needing to regain her balance as the river threatens to push her off her feet. As she starts swimming, we feed the rope out with her. She makes good progress at first, swimming diagonally against the current, but as she nears the centre of the river, she is swept downstream and the rope is pulled taut.

'Give me more slack,' she calls.

We oblige but she is just carried further along by the current. She begins to swing back towards our bank, pulled around by the rope.

'That's enough. Get out of there, Meaghan,' I yell.

As if on my cue, Meaghan is suddenly sucked underwater by an invisible undertow. Karim lets go of the rope and rushes to the river. He is about to recklessly jump in when Meaghan bobs to the surface and surges towards the bank. He supports her as she clambers out of the water, breathing heavily.

We drop the rope and rush over to make sure that Meaghan is alright, but she is none the worse for her misadventure.

Cassidy slaps Karim across the back of the head. 'Idiot. What were you going to do? Jump in and get swept away?'

He wisely decides to take the hit and say nothing.

Meaghan only takes a few minutes to recover before she insists on trying again.

The second attempt is equally unsuccessful and when she scrapes her shoulder on the third try, she concedes that the river is too choppy to swim.

'We're gonna need to build a bridge,' Poh says.

'That's gonna be tricky. How do you make a bridge with access only to one side of the river?' Cassidy asks.

'Leave the design to me. I'll make it work,' Poh says confidently. 'What tools do we have?'

Karim leads the way to the wheelchair and shows us what he and Cassidy managed to grab. We have a large axe and a hand axe, a hammer, a large handful of long nails and a small box of short nails.

'Is that it?' Poh asks sceptically.

'And a trenching tool,' Cassidy says. 'And the rope, of course.'

Poh sits down to plan a design while Smithy grabs the axe and walks towards the river.

'What are you doing?' I ask.

He glances back over his shoulder. 'I have an idea.'

Smithy finds one of the largest trees on the riverbank and starts hacking into the trunk with the axe. He is enthusiastic but a bit awkward and ineffective with only one arm. Newton and I take over from him and by lunch, we're halfway through the tree.

'I'm starving. Let's see what's for eating,' Smithy says.

'Is it safe to leave it like that?' I ask.

'Sure. I want it to fall anyway, so if it falls, it falls.' He grins mischievously and walks towards the others.

Over lunch, Poh outlines her plan for the bridge.

'So basically, it will just be like a large ladder. We cut down four trees and use the top seven metres so that it is not too heavy. We lash the trees together to make two long poles and nail planks between them to make a bridge. Simple.'

'But how do we then get it across the river?' Cassidy asks.

'Leave that to me. But I will need the rope, and another the same length made from creepers. And because I need the rope, we will need to lash the trees together using creepers too.'

'Will they be strong enough?' Karim queries.

'They'll have to be,' Poh says. 'Lastly, whoever is nailing the planks on, do not hammer them all the way in. We don't have enough nails to leave them behind, so we'll have to retrieve them and take them with us.'

We all nod our understanding and after lunch, set to making Poh's plan a reality. First, we have to finish Smithy's plan, because there aren't enough tools to go around. We continue to chop into the large, riverside tree for about an hour until, slowly but surely, it begins to topple. As it gathers momentum, there is a hideous creaking and cracking sound as it tears off from its stump. It crashes to the ground, just spanning the river and landing with its top branches on the opposite bank.

'Ah ha!' Smithy says triumphantly.

But as soon as the sound is out of his mouth the half of the tree that is submerged in the river, is tugged downstream. Depressingly slowly, as there is nothing that we can do about it, the top of the tree is pulled around and into the river.

Finally, enough of the tree is in contact with the rushing water to accelerate its movement downstream, and the entire pine casually floats away from us, looking as majestic as a tall ship on its maiden voyage.

'Unlucky,' Poh says from behind us.

'It was worth a try,' Smithy says a little defensively.

We spend a day walking upstream to try and find somewhere better to swim or cross the river, but everything is much of a muchness. The following three days, we slowly work towards building Poh's bridge. We are limited mainly by the tools at our disposal and so, if we're not chopping wood, we use the rest of our time on other jobs.

We improve our basket weaving skills and soon have lots that we can use for foraging and storing goods. We spend time foraging for creepers and food. Throughout the forest, we have noticed parasitic plants that stifle the growth of pines, without killing them completely. As we strip some of the creepers from one of these trees, I get a better understanding of how it drills into the pine to sap its energy and nutrients. I spearhead tying lots of creepers together to make three thirty-metre-long cords. Cassidy then plaits them together in an attempt to make a rope strong enough to take the weight of the bridge.

We manage to build up our food stores as well as expand the variety in our pantry. Many mushrooms have popped up since the storm and we harvest fresh ones and throw away the old specimens we have been carrying since camp. We stick to the fungi that Ayesha identified as safe, including the huge, original, orange mushrooms, but I can't help but feel that we're rejecting a whole feast of edible fungi. The danger, of course, is that lying amongst the feast is a mushroom that will make us very sick, or worse, kill us.

Meaghan perfects the art of halving tree ferns to access the tender heart of the plant. She curls one half away from the other, hopefully allowing the ferns to survive and regrow. She also digs up the roots of the smaller ferns that grow around the bases of the riverside pines. Furthermore, she finds gigantic earthworms amongst the fern roots. Some of them are at least forty or fifty centimetres long, their elongated bodies a pinkish brown colour. I offer to be the first to try them. Cut into small segments and fried they have a welcome meaty flavour, but they are also a bit gritty.

The fern roots are black and woody, and after roasting them over a fire, Meaghan goes through the procedure and eventually reports that they taste good enough. Once we know they're safe, we all hoe into the roots. They taste a bit nutty and are starchy like a potato.

While Cassidy and Newton focus on felling another pine, the rest of us dig up as many fern roots as possible. Meaghan dries some in the sun, and after roasting them to dry them further, crushes them between two stones to make a flour. Following her lead, we manage to crush and bag about a kilo of flour, reusing old food packs that we have held onto. Hopefully, this flour can last longer than the fresh mushrooms and plants that we have been relying on.

Expanding on the idea, I also manage to dry mushrooms on the sunny riverbank. When added to boiling water, they rehydrate well enough, and maintain most of their flavour.

As Poh and I are cutting down the third pine, an odd sensation creeps through my body. An uneasiness that I can't grasp but lingers in my subconscious. As I watch the large pine tree crash to the forest floor, the feeling solidifies into thoughts of regret and hesitation.

While humans screwed up the world and ultimately paid with their lives, these trees survived. They survived when all the animals and insects died, ending the reign of the flowering trees that relied on them for pollination and dispersal of their seeds.

These pine trees have stood the test of time, and now us humans have risen from the ashes, and the first thing we do is tear the trees right back down. Maybe it would've been better if the R.E.Generation Project was never formed. Maybe it would've been better if all our pods had failed and the earth could've continued healing without the destructive influence of humans ruining everything once again.

We give the axe to Meaghan and Smithy who start work on the final tree. Poh grabs the hand axe and begins to cleave the upper branches from the trunk of the pine.

'Oh, Sarah. I wanted to tell you. I've got my period,' she tells me.

'Okay.' I reply, caught off guard. 'Do you have everything you need?'

'Yeah. Cassidy gave me some of her re-useable products. It's all good. But I've got my period,' she repeats pointedly.

I stare back at her, realising I'm missing the point.

'Which means I'm not pregnant!' she says emphatically.

'That's amazing!' I say, catching on. 'What a relief.'

'Yup. Declan can fuck right off.'

'Poh!' I say taken aback. 'I didn't think that you swore.'

'Not usually, but I think that rapist deserves some expletives,' she says with a fearless grin.

I couldn't agree more.

The first two pine trees have had their tops cleared, been cut down to length and lashed together using some creepers.

By the third morning of construction, the final tree has been stripped and is ready to be shortened. Newton hacks off the top few metres, which are too flimsy, and then turns to me with the axe over his shoulder.

'How are we supposed to measure seven metres?' he asks. 'Have you got a tape measure handy?'

'It's just a guess, Newton.' I reply. 'I don't think it matters that much. I saw Karim pacing it out yesterday.'

Newton takes seven large steps along the length of the trunk and starts hacking into the wood with the axe. His muscles bulge and he's soon covered in sweat.

'My turn,' I say, holding my hand out to request the axe.

I chop into the wood, trying to alternate the angle of my cuts like the woodcutters I used to love watching at the agricultural shows. It's tiring work, and after making some progress I hand the axe back to Newton. He finishes the job quickly and we pick up the finished pole together and carry it across to the others.

There is a buzz around camp as the bridge nears completion

Karim has been busy halving some of the larger branches to make rough planks. Meaghan and Smithy help him hammer them into the poles, leaving about a centimetre of nail exposed so that we can retrieve them. Occasionally we hear loud, frustrated sounds from them as someone bends one of the precious nails, mangling them beyond use. More often than not Smithy is the culprit, as he adjusts to using his non-dominant right hand. The final product is a large, ladder-like structure, with two thick side rails made from the pine poles, and wooden planks acting as ladder rungs.

Meaghan has chosen the narrowest part of the river for the bridge's location. The bank on either side is slightly raised, allowing the bridge to lie flat and clear from the rushing water.

The previous day, Meaghan had instructed Newton to scale a tree about a metre from the bank and, using the hand axe, he trimmed the branches off one side up to about twelve metres, leaving the tall trunk exposed.

I help Cassidy check that our creeper rope is as long as the cord that we brought from camp. Satisfied, we take them both to Poh, who nods her thanks. I rush back to help Karim, Meaghan and Smithy carry the bridge ladder, but it's too heavy for just the four of us. We call Cassidy and Newton across to help us carry it carefully to the crossing point,

'Where do you want this?' Karim asks Poh.

'Lie it down here with one foot resting beside this tree and the ladder going away from the river.' She gestures to the tree that

Newton has cleared. 'Make sure the planks are face down, against the ground.'

She makes some minor adjustments to make sure she's happy with its position.

'Do you guys have any spare creeper rope?' she asks.

'Sure,' I say grabbing some for her.

Poh loops the extra rope around the foot of the ladder and the tree, loosely binding them while leaving some room for movement.

We all take a quick lunch break. Aravis and Tumnus decide that it's their lunchtime too and pester us by climbing all over us. They don't stop until I take a break to give them some scraps of mushroom and worms. They have already doubled in length and have started to accept attention from the others without threatening to bite their fingers off.

After lunch, we return to the bridge and Poh attaches the two long ropes to the top of the ladder. Hopefully, the one made of creeper can match the cord-rope's strength and durability.

She hands the ropes to Newton, along with a few plastic food wrappers, and instructs him to climb the tree. We all watch as he skilfully scales the tree, feeding the ropes out behind him and successfully navigating a way up even though he only has half the branches to work with.

'Now put the ropes over that branch, with the loose end hanging over the far side, facing the river,' Poh calls out to him, pointing to a large branch overhanging the area that he cleared earlier. 'If you can also put the plastic between the branch and the creeper-rope it'll reduce the friction.'

He follows her instructions and the two ropes hit the forest floor. They now run from the top of the ladder, diagonally up towards Newton, cross over the branch and fall vertically towards the base of the ladder.

'Alright, good to go, team,' Poh says excitedly.

Meaghan and Cassidy walk to the head of the ladder. Their job is to lift the ladder and then support it while walking along the length of the ladder as it rises.

'Just make sure you're ready to jump out of the way if the ropes or branch snap,' Karim says. 'You're both far more important than this bridge, okay. No heroics.'

Newton's job is to stay in the tree and report on the condition of the ropes and branch.

Karim, Smithy, Poh and I take up the ends of the ropes and, following Poh's lead, begin to pull the ladder off the ground. It's heavy work, but once it's off the ground, we make steady progress.

'Looking good,' Newton calls. 'The branch is holding nicely.'

We walk slowly away from the river as the giant structure slowly rises into the air. It becomes gradually easier as it becomes more and more upright, and the distance between the top of the ladder and the branch closes. Meaghan and Cassidy join us on the ropes, and with their extra efforts, we manage to get the ladder completely upright. It stands erect, directly under the branch, supported by the rope above it and the six of us braced at the other end.

'Bugger!' Newton calls. I see something float down to the ground from the branch. 'The plastic has just come loose. There's no way I can get more underneath it with the rope under strain.'

'We'll have to hope it doesn't matter,' Poh calls back. 'Everyone ready?' she asks us.

We nod.

We begin to lower the ladder down, over the awaiting river. We walk slowly forward as a cohesive unit, and it carefully drops lower and lower.

'The creeper rope is struggling,' Newton calls to us.

We stop moving as one.

'One creeper has snapped,' he continues.

'Okay. Keep us updated,' Meaghan calls back. 'Let's keep going.'

We start walking forward, but Newton calls out once again. 'It's snagged on one of the knots in the creeper rope.'

We have already taken another step before we stop and the ladder begins to tip. The cord rope has continued to lower one side of the ladder, while the creeper rope has caught and now holds more of the ladder's weight. We strain against it, gritting our teeth and trying to maintain control.

'It's snapping,' Newton calls and I feel the creeper rope in my hands lose tension as it tears in two.

Now that the creeper isn't taking any of the weight, the ladder starts to tip the other way and Poh, Karim and Cassidy, who are holding the cord rope, are jerked forward as they take the ladder's full weight.

I watch in horror, unable to do anything to help as our construction plummets towards the river, dragging my friends with it.

KARL

Karl hums to himself as his car drives him towards work. Despite all the doom and gloom on the news, he is happy today. As of midnight last night, Denmark has left the European Union and United Nations. They are no longer bound by the Hunger or Displaced Persons Acts.

For years, Denmark has been sending more than a third of its produced food to other countries. Feeding those who can't feed themselves. And what did they get in return? Refugees and unskilled workers. More mouths to feed and people to house.

But no more. From today, all the food that Karl and his company *Origin Food* produce, will go directly to the Danish people. Advances in technology have allowed Denmark to become self-sufficient. *Origin Food* owns huge buildings crammed with hydroponic plants of all varieties. Broad ecosystems, complete with insect pollinators and UV lighting. Kept safe from the extreme weather, these factories are able to produce all year round. Other companies farm animals indoors. Multi story buildings full of pork, beef and poultry. The emissions are captured and turned into fuel. The footprint of agriculture has been reduced monumentally, and Denmark's food production has been safeguarded for the future.

Karl's car pulls into the enclosed parking lot and large double doors close behind him, allowing the air temperature to be easily controlled. Karl steps out of his car.

'Thanks, pal, Lock doors,' he says.

The car responds as directed, recognising his voice signature.

'Hey, Freja,' Karl says, passing reception.

'Karl,' Freja replies. She is a competent receptionist but prone to gossip and wild theories. 'Have you seen the weather forecast?' she asks.

'Yes,' he responds with a sigh. 'Another storm. Nothing we haven't seen before.'

'They say this one could be the worst yet. The UK has already had two straight days of rain.'

'Sod 'em,' he says unsympathetically. 'We'll be fine. Now get back to work.'

He walks up to his office and syncs with the computer. He lies down on the couch, settling in for another day of spreadsheets, data and phone calls.

They've got robots and AI systems that could do both Karl and Freja's jobs, but luckily, the Danish government has Minimum Human Employment laws.

Stuck in his central office and plugged into the system, Karl has no contact with the weather outside. He doesn't hear the raging wind or feel the torrential rain. But when the power goes out, even he can't ignore it.

'Come on, come on,' he mutters, sitting up.

He waits patiently and a few minutes later, everything buzzes back into life. He checks to see where the power is coming from. The backup batteries have activated. They have enough power to supply the whole plant for a week. Surely the power will have been reconnected by then.

Karl completes his day's work and heads down to reception.

'Where do you think you're going?' Freja asks.

'Home,' Karl replies.

'Do you pay any attention to the news?' Freja inquires.

'Unlike you, when I am at work, I am working. Not watching news and scrolling social media.'

'It's good to stay informed. For example, I know that the roads are flooded. We're totally cut off.'

'Seriously?' Karl asks incredulously.

'Yup, we're going to have to stay here, I guess. Just you, me and Stig.'

Stig is the only other worker at the food production plant. He services the machines and robots and makes sure that everything runs smoothly growing the vegetables.

'Oh joy,' Karl says and turns his back on Freja.

He retreats to his office and pulls an instant meal out of the cupboard. Chicken Tikka-Masala. He shakes the box to warm up the contents before digging in. Halfway through his meal he accepts a call from his boss.

'Karl. My main man. How are things holding up over there?'

'Fine, Ida,' he replies. 'We lost power, but the batteries have kicked in. Only problem is the floods have us trapped here.'

'It's the same everywhere,' Ida says. 'The winds and floods have affected power plants and power lines all over the country. Only one of our plants is still connected. The bigger problem for us is that the hail has decimated our outdoor fruit trees. Ten-percent of our total expected production wiped out in a day.'

Karl raises an eyebrow in surprise. 'I'm sorry to hear that, Boss.'

'I just wanted to make sure that everything was fine with you. We definitely can't afford to lose any yield from your crops,' Ida states bluntly.

'Everything is fine, Boss. No need to worry about us.'

'Great. Say hello to Stig for me,' Ida says.

'Will do. Speak soon.' He disconnects the call and shakes his head. The day had started so promisingly too.

Stig and Freja come and visit him, but he sends them away, preferring his own company to theirs. He sleeps on the couch and wakes stiff and sore.

'Lights on,' he commands, but nothing happens. 'I said, *lights on*.'

He stands up angrily. Without power, he can't check what has happened. He will have to check the batteries manually.

He exits his office and heads for reception, following the self-powered safety lights that illuminate the corridors. They glow orange at ground level, with just enough light to see, but they create an eerie feeling through the whole building.

Karl finds Stig and Freja sleeping on the floor behind Freja's desk.

'Wake up, guys, get moving,' he says loudly.

'What time is it?' Freja asks.

'Lights on,' Stig calls.

'You think I didn't try that already?' Karl says angrily. 'Come on, Stig, we need to check on the batteries.'

Stig groans as he pulls himself to his feet.

'With a bit of enthusiasm,' Karl barks patronisingly. 'You do realise that this production plant is supposed to create enough food to feed fifty thousand people.'

'I know, I know,' Stig grumbles.

The batteries are stored in the basement, and as they descend the stairwell, they realise that the underground levels are completely flooded and they can't get to the bottom of the stairs.

'You need to get to the batteries,' Karl commands.

'No way,' Stig replies. 'It's way too dangerous.'

'The batteries will still have charge,' Karl says optimistically. 'We need to get them out of the water, dry them off and reconnect them upstairs.'

'We don't know if the water level is still rising and dealing with batteries in water is dangerous,' Stig explains.

'But you're the engineer. I need you to get them.'

'I'm the engineer, and I say it's too dangerous,' Stig says firmly and starts walking back up the stairs.

Karl retreats to his office and finds an old handheld telephone in his bottom draw. Old school. Luckily it still has some charge left.

He calls Ida and updates her on the situation.

'Shit,' she says bluntly. 'I've just been on the phone with Malthe. Apparently, they have a leak that is threatening a section of their plant. He's hopeful the robots will be able to repair it though.'

'That's the problem,' Karl states. 'Without power, we don't have robots here.'

'Keep me updated,' she says curtly and cuts the line.

The flood waters don't recede and Karl, Freja and Stig are stuck at the plant for the next three days. Karl takes to walking amongst the plants. He visits parts of the facility that he has never been to before. Without power to provide light or fresh nutrients to the plants, they start to wilt and die. On the third day, as Karl walks his lap of the factory, he steps around a few capsicums that have fallen off their plants and onto the floor.

'We need to get those batteries,' he says to Stig when he gets back to reception.

'Go for your life, Karl,' Stig replies, staying in his chair and crossing his arms.

'Wimp,' Karl sneers and heads for the basement. When he reaches the water on the stairs, he ploughs into it, continuing to descend until it is around his chest.

'See, it's not too bad,' he says to himself.

He wades through the brown water, using the torch on the handheld phone to guide his way. The safety lights are under the water causing patches of water to glow eerily. He finds the battery room and pulls the door open. Inside are six battery units. Even two would make a huge difference if he can get them upstairs.

He approaches the first unit and appraises the portion that remains above water. He takes a deep breath and ducks below the surface, using the torch to assess the battery. He finds all the connection points and painstakingly disconnects them, returning to the surface to breath as needed. Once the unit is disconnected from the wires and unlocked from the floor, he drags it out of the room and towards the stairwell. He hauls it up the stairs with difficulty, the unit becoming heavier and heavier as more of it is exposed above the water. He leaves the unit on the first landing and plunges back into the water. He is starting to get cold, but he should be able to get another unit before he needs to dry off and warm up. If Stig was here, they could work twice as fast, he thinks bitterly.

He enters the battery room and starts disconnecting the second unit. Emboldened by his earlier success, he speeds up the process, taking less care with each movement. One of the wires from the first unit wraps itself around Karl's leg without him noticing. As he grabs the last wire to disconnect it from the second unit, current flows through his body The huge electric shock interferes with the signals in his heart, causing it to stop instantly.

The death of one selfish man, inconsequential in itself. However, the repercussions will be catastrophic for thousands of people who rely on the food that he provides for their survival.

SARAH

CHAPTER 13

'Let go of the ropes, let go of the ropes,' Newton calls in desperation, hoping to save the ladder from swinging too violently and landing in the river.

Cassidy, Karim and Poh let the cord rope go and it whips away as the ladder comes crashing to the ground.

'Ow. Fuck me,' Karim yells.

'What happened?' Poh asks.

'The rope got me as it whizzed past,' he says holding a hand to his face tenderly.

A long raised mark has already started to emerge, running from his jawline to his eyebrow.

'Come here,' Cassidy says warmly. 'I'll check you out.'

She leads Karim back to our temporary camp.

'Oopsie,' Smithy says.

'At least the bridge seems to be fine,' Meaghan says.

We all turn to look at the ladder that now straddles the two banks of the river and provides a simple bridge across the raging water. The area where the pines were lashed together appears to be holding, despite its rapid descent and jarring collision with the ground.

Newton climbs down the tree to join the rest of us. 'Lucky you tied it to this tree Poh. Otherwise the bottom might have bucked or slid all over the place.'

She winks at him knowingly. 'That was the general idea. Although I would've preferred it to not fall at all.'

'Who's gonna be the first to try it out?' Smithy asks excitedly.

'I'm happy to,' Meaghan says raising her hand. 'Unless you want to Poh? It's your bridge, after all.'

'No. I am more than happy not to go first,' Poh replies, looking at the bridge nervously.

Meaghan approaches the bridge cautiously, before nimbly stepping onto the first plank and walking out over the water. The bridge dips slightly in the middle as she crosses but holds firm as she glides across to the other side.

'Feels fine!' she calls from the far side, giving us a big thumbs up.

'But the less time we spend on it, the better,' Poh says quietly to the rest of us.

We call instructions across to Meaghan who unties the cord-rope from the bridge and secures it to a large tree on the opposite bank. We wind the other end around a tree on our side of the river, before tying it off, making a handrail above the bridge.

We load up our backpacks and prepare to cross, with Poh shouldering Meaghan's bag. We each take some of the extra items from the wheelchair, either managing to stuff them into our bulging packs or carrying them in our hands.

Cassidy tells us that Karim's face should be fine. He has a raised welt the length of his face and his right eye is swollen, but no major damage done.

We walk to the bridge together, pushing the empty wheelchair. Although the long creeper rope snapped in half, we still have about fifteen metres of rope on our side of the river. Poh ties this to the wheelchair as Smithy makes his way across the bridge.

He is a bit unsteady, his balance impacted by the loss of his arm. I watch on nervously, holding my breath until he is safely on the other side.

Newton and Cassidy take turns to carefully walk across the bridge, holding onto the rope for balance. Each of them also carries a pouch with a lizard inside.

Poh hands Karim the loose end of the creeper rope as he approaches the bridge.

'Take this to the other side, but don't pull on it until I tell you,' she says.

He takes it and tenderly crosses the bridge, carefully trailing the rope behind him and making sure it doesn't snag. Once on the other side he turns and looks at Poh inquisitively, but she holds up her hand, telling him to wait.

She unties the cord rope from the tree, passes it through the front triangle of the wheelchair and ties it back onto the tree making it as taught as possible.

'What's your plan?' I ask her, intrigued.

'To hope like hell that this rope is strong enough?' she replies. She pushes the chair to the river's edge, and waves to Karim. Newton and Meaghan join Karim on the other end of the creeper rope and they pull hard. The chair jerks off the riverbank but stays suspended in the air thanks to the cord rope. By hauling on the creeper rope the others are able to drag the chair to the other side.

Poh turns to me, a nervous look on her face. 'Do you mind going last and pulling up the planks?' she asks.

I look down at the hammer that's already in my hand. 'No worries. Go for it.'

Once the others have hauled the chair out of the water, she unties the rope once more and ties it around my waist. She walks up to the bridge, takes a deep breath and then starts to crawl across, choosing to stay on her hands and knees.

Her arms shake and she makes the task more difficult for herself as her entire body is incredibly tense. I realise that she's been putting off crossing the bridge herself, but now that she's on it, she manages to keep moving, bravely overcoming her fear and powering on.

She reaches the other side and turns back to face me with a nervous grin, still shaking from the adrenaline.

I crawl across the bridge too, going backwards and removing the nails from the planks as I go. The pack on my back keeps trying to fall to one side and unbalance me, but I manage to keep it under control. I leave the planks behind and pocket the long nails. Every fourth plank I leave nailed in to make sure the bridge maintains its structural integrity. The last thing I want is for it to collapse while I'm crossing.

It's a painstaking task, but the others call encouragement to me as I approach and I fall into the rhythm of the task, finding the optimal way to reverse over the planks and to pull out the nails.

Once I'm on the other side, we stand in a circle and everyone grins at each other.

'Not sure about you guys,' Smithy says. 'But as much as I've loved this river, I'm ready to leave it behind.'

'Here, here,' Meaghan says enthusiastically.

I smile, happy to see the two of them getting along.

We make sure all of our canteens are filled with water, load up the wheelchair with all our excess baggage and continue our march south-west. The ground becomes very hilly and we spend the afternoon climbing small rises, just to descend the other side and be faced with a new hill. Fortunately, as it nears dinner time, we hit a consistent downhill stretch. We follow it mindlessly, and the pine trees begin to thin out. I look up, and through the last of the trees, I see the blue-green waters of a gorgeous lake.

The pines stop a few metres from the water and are replaced by knee high grasses that sway in the afternoon breeze. The pine trees begin again on the opposite bank, covering a steep hill. The far shore doesn't look all that far away, but the lake stretches to a close horizon on either side, disappearing behind hilly promontories and tree covered fingers that jut into the lake.

'Now we turn left and go clockwise around the lake, after which we follow one of the rivers that feeds it on the north edge. Well, I guess I should say south edge,' Smithy explains.

'Wouldn't it be quicker to swim across?' I ask. 'It can't be that far.'

'If it's similar to what it was before it's almost two kilometres,' Smithy says.

'I'm not swimming two k's,' Poh says.

'I can't swim two k's,' Smithy adds, waving his short stump.

'Neither can I,' Karim says. Cassidy turns to look at him. 'I can't swim that far,' he continues defensively. 'I can manage fifty metres if I have to, but anything more than that, no chance.'

'How is that possible?' Cassidy asks incredulously.

'How were you chosen for surviving the end of the world if you can't swim?' Meaghan adds.

'I don't think it was a prerequisite,' he says. 'They were more interested in my degrees in electrical engineering and accounting.'

'Well at least you're not as boring as I always thought,' I say teasingly.

'Anyway, the swim is a no-go,' Smithy says, getting us back on track. 'I say we camp here for the night and start out again in the morning.'

After a long, stressful day crossing the river, no one is going to argue with that logic.

While the guys gather firewood and build a fire to cook dinner, Cassidy, Poh, Meaghan and I plunge into the lake to freshen up.

The water is warm, and we churn up the bottom with every step, turning the water brown and murky. Cassidy and Meaghan undo their ponytails and shake their hair out. Cassidy's long, blonde hair darkens as it hits the water and splays out around her. Meaghan dives under the water and after a few seconds, I start to panic that she hasn't resurfaced. I scan the murky water hoping that she didn't hit her head. Suddenly, a hand clamps around my ankle and she surfaces next to me.

'Jesus! You scared the crap out of me,' I exclaim.

'I didn't know that you were religious,' Meaghan says.

'I'm not. Not anymore, anyway. Shut up, you.'

She pokes her tongue out at me and dives under the water again.

'What's got into her?' Poh asks.

I shake my head. 'No idea.'

Meaghan pops up besides Cassidy who splashes water in her face.

'Hey!' Meaghan says, splashing Cassidy back.

I'm showered by a spray of water and turn to see Poh grinning at me cheekily.

'Oh, you're on,' I say. I dive underwater and scoop a handful of water towards her as I stand up.

We continue to act like children, splashing and tackling each other in the shallow waters of the lake. None of us have laughed this much since we woke up. I glance at Meaghan and she winks at me, as if reading my mind. This was exactly what she was trying to do.

Poh is hunched over double, trying to get her laughter under control. She makes a succession of small hiccup-like sounds.

'Hey, guys, do you think we need to worry about crocodiles?' Cassidy says suddenly.

'Nah, we're way too far south,' Meaghan says. 'I mean north. I mean we're too far away from the tropics to worry about that.'

'The weather is pretty tropical here nowadays,' I point out.

'And the water is pretty warm,' Cassidy adds.

We all rush to the shore, continuing to laugh stupidly as we go.

I'm not convinced that there are any crocodiles, but now that Cassidy has mentioned it, I'm unable to get the thought out of my head.

Meaghan shakes out her hair and it returns to its bright, scarlet colour as it dries. 'I wonder what the boys have cooked for dinner.'

'Something smells good,' Poh says, raising her nose to the air.

We walk barefoot through the grass towards the fire and the smell of cooking meat. The boys have set up a few metres back into the pine trees.

'You guys need to wear more clothes–' Karim starts to say, but Poh is quick to cut him off.

'We'll wear whatever the hell we like, thank you very much. No man will tell us what we can and can't wear.' She puffs her chest out to emphasize the point.

'I was just saying you should wear more clothes, like shoes and pants, if you're walking through the grassy area, because there's snakes in there,' Karim replies.

'Oh,' Poh says, glancing over her shoulder at the grass. 'Makes sense.'

'We killed two different types and collected a variety of seed heads too,' Smithy says.

They have lain out the newly foraged foods on different plates. We each choose a plate to be the tester for these new foods. Everyone wants to try the snake meat, but we let Smithy and Newton have it because they found and killed them.

'My grass seeds taste like grass,' I say sadly.

'Mine too,' Meaghan agrees.

'I think mine would be better roasted,' Cassidy says.

'I hope you didn't think that was all we were giving you for dinner,' Karim says with a grin.

He pulls a pot out of the fire and dishes a stew onto everyone's plates. 'It's mushroom, pine seeds and flour-balls,' he says proudly.

I scoop some of the stew into my mouth and smile broadly. 'It's delicious!'

'I know,' Karim says proudly. 'I was surprised too. I think it's the bits of worm that add a meaty flavour. I like to think it's bacon.'

Poh spits the stew back out onto her plate. 'But I haven't eaten worm yet. I don't think I want to,' she says.

'But it does taste delicious,' Cassidy says.

Poh rolls the saliva around in her mouth, tasting the remnants of her rejected mouthful. 'Yeah, it kinda does,' she says sheepishly.

She eats a small mouthful, chewing tentatively, followed by another, and another. Soon, she is wolfing it down like the rest of us.

The group is well fed and relaxed, and we spend the evening talking about nothing and everything. As it gets later, we lie down and curl up under our blankets around the fire. We keep chatting as the world descends into darkness around us. One by one, people stop replying as they drift into sleep, until it is just Karim and I left awake.

He whispers, to not wake Cassidy who is nestled up beside him. 'You know, you've done a wonderful job.'

'I haven't done anything,' I reply.

'Bullshit. Look at all this.' I see his outline gesture to the temporary campsite and the slumbering bodies. 'This is a group of people who are happy and feel safe and content. At this point in time, they don't have a worry in the world, and that's pretty impressive for a bunch of randoms, thrown together and asked to survive after the world has ended. We have literally all of the troubles in the world. What can we eat, where can we sleep, when am I going to die? But you keep a level head. You keep everyone grounded and everyone safe. Don't underestimate the power of a loving leader. You've done a wonderful job.'

'Nice speech, douchebag,' I reply, but I'm grinning from ear to ear, lapping up his praise. Deep down, I know that I'm doing a good job. I care about these people and all I want is for our mission to be a success. However, even when you know you're doing a good job it feels wonderful to hear someone else say it. No matter how confident we are, we all need validation.

I make a note to tell the others, more often, what they're doing well. I want them to feel what I just felt thanks to Karim.

'I guess I'd better take first watch,' I whisper to him.

He makes an indistinct noise of agreement.

'I'll wake you in a few hours to take over.'

Grunt.

I'm not convinced that he hasn't already fallen asleep.

The night is not cold, but I wrap my blanket around my shoulders and walk a lap around camp. We've had no night-time incidents since we woke up, but we're moving into new territory with no knowledge of what danger might lie ahead. Even so, it's hard to concentrate and not let your mind wander when you're not expecting anything to happen.

As I approach the lake, the line of pine trees recedes, and the sky opens up above me. There is no moon tonight, but the stars provide all the light I need to see the way.

I walk to the lake's edge and stare into the heavens, trying to find something familiar. I'm not used to seeing so many stars, and they appear to make a dome high above my head. Instead of feeling thousands of light years away, they seem like they've been painted on the ceiling of our world. This is what the stars must've looked like to ancient civilisations, before light pollution and computer screens. When people looked up, instead of down.

I can understand why people believed that God or the gods came from the heavens. It's breathtaking. Mesmerizing.

The Milky Way stretches across the sky, a marvellous band of millions of stars, sharing their light with my tiny eyes. After a few minutes, I find the southern cross. It's not as clear as I am used to, because there are many smaller stars dotted within it that I've never seen before. Yet there she is, the exact same constellation that my father taught me to find, using the pointers to ensure that I'd found the real one. He also taught me how to use it to find south, which is now north. Maybe I should be calling it the northern cross now, as we're technically in the northern hemisphere. Suck on that, Europe, America and Asia. Guess who's on top of the world now?

The beautiful and simple southern cross. The exact same stars that the ancient Incas saw and that will probably still be shining in millions, if not billions of years. It gives me cosmic vertigo to think about how long they'll be a constant in the night sky. It reminds me how much humans are just a flash in the pan as far as the universe is concerned.

Gazing at the billions of stars above me I can't help but feel that there must be some other life out there. They always said that the biggest difficulty in finding them is the huge distances between stars. But there are countless stars within thousands of light years. It's now been thousands of years since humans last inhabited this planet. They used to broadcast messages and maybe now, enough time has passed that a distant civilisation has received and deciphered those messages. Maybe our story ends with extra-terrestrials arriving to save or colonize us. The latter seems more likely, if they are anything like humans.

I spend the next few hours walking laps around the camp and stopping at the lake's edge to stargaze. I wake Karim and curl up under my blanket. Even though I'm weary from the long day, and the rigours of our new lives, it takes me a while to fall asleep as my mind tries, and fails, to grasp the vastness and sheer magnitude of the universe.

NGIDI

2100 was a good year, the year the world finally got its shit together and ended world hunger. They put systems in place so that the food was dispersed evenly. Somehow, people stopped listening to money for a few consecutive years and realised that the haves could feed the have-nots.

Ngidi shakes his head. He wasn't old enough to remember those good days. And when food started to run out again, who did the world cast adrift first? The Africans. Last in, first out. Something like that. The biggest problem was that the governments in Africa had become complacent and expected the support to keep coming. Instead of investing in greater food production, they invested in technology and business. Economies boomed, and people started starving as soon as they were cut off.

Ngidi waits in line at the commissary, his weekly tokens grasped firmly in the palm of his hand. He doesn't want to risk having them stolen. Only last month, the woman in front of him got to the front of the queue to find that she had been pick pocketed and wouldn't be taking any food home for her family. Of course, she blamed Ngidi at first, but he was able to convince her that if he had stolen her tokens, he wouldn't have stayed standing behind her.

The line shuffles forward and Ngidi smiles at the man who has just received his provisions. The man smiles back, but there is no joy in his eyes. Ngidi turns to watch him leave and sees a sore on the back of the man's neck. The skin looks necrotic and Ngidi notices movement within the wound. Ngidi manages to stop himself from hurling as he realises that there are maggots living inside the festering wound.

Ngidi reaches the front of the queue and smiles to Akuchi behind the counter.

He hands over his tokens and she retrieves a large sack for him. It seems like a lot, but when it needs to last a week and is divided amongst seven people, it barely touches the edges of their hunger.

'Any chance of a bonus this week?' Ngidi asks.

She shakes her head. 'I'm not sure bonuses are a thing anymore.'

'How the hell are we going to survive, then?' he asks, frustrated.

She shrugs.

'I have three children, a wife and my parents to feed at home. Surely, I should get more than smaller families,' he says in exasperation.

'You know the rules. Same provision per family. I didn't make them and there is nothing I can do about it.'

'How is that fair?'

'It's not. But they tried to disperse food per person, but then people lied about how many family members they had to get more. Plus, I think that they are hoping that the 'food per family' policy encourages people to have fewer children. An unofficial one child policy.'

'That's all well and good for new families, but it doesn't exactly help those that already have children.'

Akuchi shrugs again. 'What can I say? Systems always seem great in theory, but never work in practice.'

'This is shit,' Ngidi says dejectedly.

'I'm sorry, but you'd better move along. Everyone else needs food too.'

Ngidi looks sheepishly over his shoulder. 'Sorry,' he mouths at the other frustrated faces.

Ngidi walks home in the heat of the day. Even in winter, they hardly get any cold days anymore. He gives the sack of food to his wife Mirembe who, considering the circumstances, cooks a wonderful meal. The seven of them sit around the table long after the last mouthful is devoured, almost as if by sitting there longer, it will make them feel like they ate more.

Later that evening, Ngidi reads his children a bedtime story and kisses them on the head.

As he exits their room, he glances back at them, making sure he has their faces etched perfectly in his memory.

'Are they asleep?' Mirembe asks as she dries the dishes.

He nods and sweeps her into his arms.

'What are you doing?' she asks, almost dropping the mug that she is holding.

'I want to dance with you,' he says.

'Okay, but just let me finish these dishes first.'

Ngidi shakes his head playfully. 'No, I need to go check on my parents soon. You can finish them then. Dance with me.'

She leans in close and they dance together around the kitchen. Mirembe's head rests on his shoulder and she can hear his heart beating inside his chest.

'Are you okay, my darling?' she asks. 'Your heart is racing at a million miles an hour.'

'It's just the dancing,' Ngidi replies. 'And the lack of food.'

She nods.

'You know that I love you, don't you?' he asks.

'Of course, I do.'

'None of this changes that. I love you. I love you. I love you.'

'I love you too, darling,' she says planting a kiss on his lips. 'Now go check on your parents and I'll finish these dishes. I'll see you in bed.'

He kisses her on the forehead and leaves her to finish her task.

On the way to his parent's room he stops by his own bedroom and retrieves a stuffed sock from the back of his drawer. He tips the contents out into his hand and walks slowly towards his parents. They are both quite old, but healthy. They haven't lost any of their mental faculties and the children love them to absolute bits.

Ngidi enters their room quietly and is glad to see that they are both already asleep. He locks the door behind him and steps up beside his mother. His hands are shaking, but he knows this is the right thing to do.

He lifts the pistol and points it at his mother's skull. He pulls the trigger and is surprised at the recoil. The pistol jerks in his hand and he rubs his shoulder. He points the pistol at his father. The old man has opened his eyes and is staring directly past the pistol at Ngidi. Ngidi can see fear in his father's eyes, but also a steely resolve. He nods, understanding what Ngidi is doing and giving him permission to end his life.

A large lump forms in Ngidi's throat as emotion threatens to overwhelm him. He looks away and pulls the trigger, unable to watch the face of the man that he loves and admires more than anyone, become a bloody mess.

Ngidi slumps against the bedroom door and slides to the ground. He wants to weep, but all he feels is numb.

He can hear his wife running towards him. 'What did you do?' she screams.

Ngidi thought he was doing a brave thing. Something that had to be done so that his wife and children had enough food to eat.

Mirembe hammers on the door, trying to will herself into the room.

Ngidi realises that he isn't brave. He is a coward. It is cowardly to leave Mirembe on her own to look after the children. It is cowardly to leave her to deal with the deaths on her own.

He knows that he should open the door. Let his wife in so that they can hold each other. Cry together. Shout curses at the world together.

Instead, he lifts the pistol to his chin and pulls the trigger.

SARAH

CHAPTER 14

We spend the next two days following the shoreline of the lake, appreciating being out in the open and on consistently flat ground. By zigzagging with the edge of the lake, we cover more distance than we need to, but it ensures that we don't get lost.

Inland, the air is less hot and humid. Occasionally, a soothing breeze flutters across the water and helps to dry our sweat-drenched clothes.

Aravis is now getting too heavy to carry in the pouch around my neck, but as much as I encourage her to come out, she invariably ends up back inside at some point. When I manage to tempt her from the warm depths, she rides cautiously on my shoulder or patters along beside me. When Tumnus is also out of his pouch, they play together, chattering to each other and sometimes play fighting. Initially, I'm a little worried that she might take off into the forest, but she seems to feel safest beside me.

Tumnus is a bit braver than Aravis and spends more of his time on the ground. He tends to stay close to Cassidy but has also started to get acquainted with the rest of the group. We continue to feed them mushrooms and worms, and they seem happy enough, continuing to grow at an impressive rate.

None of us show any adverse reactions to the new grains and snake meat that we had for dinner, which means we can all eat varied and exciting meals. As we follow the shoreline, we often see large snakes sunbaking near the water's edge. Meaghan and the boys take it in turns to hunt them, and we have a steady supply of meat which nourishes our bodies and our minds. I get the impression that the snakes don't have any natural predators, because they don't flee when we approach. Some of them bear their fangs, but they are no match for a well-placed axe swing.

In the evening, we create a makeshift fishing rod using the cord-like rope and a stick. Karim and I are determined to catch something and wait patiently for a few hours before giving up.

'So, maybe there are snakes, but no fish,' Karim ponders. 'I don't get this new world…'

Late on the second day, as we trudge in silence, each in our own world as we look forward to stopping for the night, Cassidy suddenly cries out.

'Tumnus!'

I glance up and see her chase Tumnus along the water's edge, before he rapidly changes direction and races into the long grass.

As if in reply to Cassidy's shout, Aravis starts squirming inside the pouch, before poking her head out and tasting the air. She moves in short, erratic bursts as she clambers out of the pouch and jumps to the ground. Cassidy is already running through the grass, chasing after Tumnus, and I follow quickly as Aravis heads in the same direction.

Tumnus waits just inside the pine trees for Aravis to arrive. They both dart around in a seemingly random fashion, before they abruptly race off into the trees. Cassidy and I tear after them, struggling to keep up despite being many times their size. I take a moment to glance over my shoulder and see the others rushing after us, Karim bringing up the rear as he struggles with the wheelchair.

We run uphill, away from the lake for five minutes until our way is blocked by a steep cliff. Aravis and Tumnus run along the face of the cliff until they find a wide crack, about a metre off the ground, and they disappear inside.

Cassidy and I stand peering into the crevice, hands on hips as we try to catch our breath. I grasp a stitch in my side, trying to pinch the pain away.

'What on earth was that about?' I ask.

'No idea,' Cassidy pants back as the others begin to arrive. 'But I can't see them in there. It looks pretty deep.'

Newton and Karim amble up to us, Newton pulling the wheelchair from in front and Karim pushing it from behind.

'I needed a workout, so thanks for that,' Karim says with a laugh.

'Where are they?' Newton asks.

I gesture to the crack in the wall of rock. 'They disappeared inside. Looks pretty deep. No idea why though?'

'I can check what's inside if you like,' Poh says.

I look at her questioningly.

'I may be afraid of heights and crossing rivers, but small spaces are my jam,' she continues.

'Okay, if you're sure,' I say. 'We don't have any torches or anything.'

'I'm sure it'll be fine once my eyes adjust.' She drops her pack onto the ground and pulls herself up towards the crack in the wall. She disappears inside, headfirst and twisted to the side to fit through the small gap.

'Is she gonna be able to get out?' Smithy asks.

No one responds and we wait with bated breath to hear from her. Cassidy peers into the crack.

'I can't see her either,' she reports.

'What the hell?' Meaghan exclaims.

Without warning, Poh's face appears in the crack, a huge smile stretching across her face. 'You guys have got to see this. Hand me my pack.'

Newton passes it up to her and she wriggles out of sight once more.

'I guess we're going in,' I say. I climb up to the fissure in the rock face and work my way into the crack without hesitation. I'm hoping that my confidence rubs off on any of the others that are sceptical about following.

I feel my way using my hands, but after about a metre, the walls on either side open out and away from me. My eyes slowly adjust to the gloom and I realise that I'm standing in a large cavern. As I step to the side, light momentarily comes into the cave from behind me, before someone else steps into the crevice, blocking it once more. Light also enters the cave from two other points high above our heads, where cracks must stretch to the surface.

Poh is leaning against the back wall waiting for us and doing her best to camouflage with the uneven rock surface. Aravis and Tumnus are curled up together near her feet and seem to be resting peacefully. It's cute to see them cuddling together, rather than relying on us.

The others file into the cave one by one. The sky is getting dark outside, so it is getting harder and harder to make out who's who. Luckily my eyes continue to adjust to the darkness, and I think I can work out who everyone is by their body shape and posture.

'Cool cave,' I hear Karim say.

'Too bad we don't know what made Tumnus and Aravis run here helter skelter,' Cassidy adds.

'It's starting to get windy outside,' Smithy says, who was the last one through the crevice. 'Maybe there's a storm coming.'

That would make some sense.

'Then we'd better make sure we bring everything inside,' I say.

Most of us brought our packs into the cavern with us, but there are a few backpacks and baskets of supplies on the wheelchair that need to come in.

We make a human chain with Newton and myself outside, Poh in the crevice and the others inside the cave.

The sky, that has been blue all day, is now covered with a layer of clouds. The weather is definitely turning, so we rush to get everything inside as quickly as we can. I pick up Karim's bag and, as I pass it to Poh, I notice how heavy and solid it is. Whatever he has inside is definitely not food or clothes. They feel cylindrical and I wonder again what he thought was worth bringing, and where he would've got them from.

Lifting the bags and tools is enough exercise to bring my stitch back and tightness to my lungs. I pass a basket of wood to Poh and feel a hand on my shoulder.

I turn to find Newton staring at me with terror in his eyes. His breathing is short and shallow.

'I...can't...breathe,' he gasps.

Even as he is talking, I can feel my own breath becoming shallower as I struggle to get enough oxygen into my lungs.

'Into the cave,' I say.

I grab the final basket of food from the wheelchair but have to leave some tools behind. I scramble up to the crack and follow Newton inside beginning to feel lightheaded and a bit unstable on my feet.

Newton sounds horrible ahead of me, hyperventilating and unable to talk. There is a cacophony of noise as the others try and work out what's happening.

'Can't breathe,' is all I manage to get out as my own breathing reaches alarming speeds. I try to slow it down and take deep breaths, but it feels like my lungs are already full and I start to see stars as my blood becomes deoxygenated.

'My bag,' Karim says, searching for it unsuccessfully in the darkness.

Poh tosses it to him with a clunk. He fumbles with the zips in his haste but manages to get it open and pulls out a metal cylinder. He tosses it to me, and I realise it's a small oxygen tank with a mask attached.

I put it on gratefully and open the valve, feeling fresh oxygen leak into my lungs and my breathing rate slow. Karim holds another oxygen mask to Newton's face, who is slumped to the side and struggling.

'How many of those do you have?' Poh asks from near the entranceway. I realise she has started to hyperventilate as well.

'Three,' Karim replies.

'Shit,' Smithy says.

The sounds of laboured breathing fill the cavern as more of the others start having difficulties.

'We'll just have to share,' Karim says, beginning to sound breathy himself. 'And use re-breathing.'

He digs out a handful of used food packets from the bottom of his bag, and hands one to each of us.

'Take a mouthful of air, turn off the tank and then breathe into the bag, okay? Put your hand up or wave if you need the oxygen tank,' he explains.

Poh raises her hand and he passes her the third tank.

I take a deep breath, hold it in my lungs, turn the oxygen off, put my bag to my mouth and blow into it. I then keep breathing from the bag, trying to keep my breaths slow and deep, as Cassidy takes the oxygen tank from me.

Six of us need to share two tanks for the first ten minutes as Newton needs his own tank to rehabilitate. Karim tries to use his flint, but he can't create any sparks.

Heavy rain starts to pour down outside, but it is strangely muffled by the cave, and we sit huddled together in an eerie silence. We are unable to talk to each other, so the only sounds we hear are each other's breathing and the faint rainfall.

That's how we spend the next few hours. Swapping between breathing into a plastic bag and getting fresh oxygen from the tank. I think we all push our bodies to their limits as we try to conserve oxygen. I feel woozy more often than not and have bouts of sweating and confusion. I use these as my indicators that I need the oxygen tank.

Meaghan clearly pushes herself too hard, passing out at one point and requiring an extended stint on the oxygen tank.

The monotonous but life-maintaining routine is broken by a few small specks of light flaring brightly on the ground. They last less than a second before they're gone. In the darkness of the cave, their light is blinding, and after they've burnt out, I have colourful dots burned into my retinas.

Karim tries to spark his steel and a few smouldering pin pricks fall to the floor.

I am about to take the bag away from my mouth when he speaks.

'Not safe yet. Oxygen please.'

Although we've already been waiting hours, the wait over the next few minutes feels the longest yet.

Karim regularly tries his flint and steel, and eventually a cascade of bright sparks falls to the ground. I wait for confirmation from him, and a minute later he tells us that it's fine.

I take the food packet away from my mouth and lie down exhausted.

'Are you okay?' Poh mumbles.

'Fine, fine,' I reply.

No one talks much as we all recover and simply appreciate the extended amount of fresh, oxygenated air. Someone moves around the cave, but most of us dwell in our own little worlds, digesting how close we came to death.

'Meaghan and Newton are fine,' Cassidy says from the darkness. 'Everyone else okay?'

We mutter agreement.

'Smithy?' Cassidy asks anxiously.

'I said I'm fine,' he replies.

'Sorry, I couldn't tell.'

'Well, that was scary,' Poh says, voicing everyone's thoughts.

'That's an understatement,' Meaghan gasps, still sounding worse for wear.

'Thanks, Karim,' Smithy says.

'Yeah, thanks mate,' Newton adds in a hoarse voice.

I crawl across the floor to Karim and lean in close to him.

'You did a wonderful job,' I whisper in his ear. 'You're the hero now.'

'Thanks, Sair,' he mumbles back.

We make a small fire using a basket of wood and kindling that we collected as we hiked.

'Are you sure that's a good idea?' Cassidy asks. 'Won't it eat up oxygen?'

'As long as there's enough oxygen around, it should just suck more oxygen into the cave,' Karim says. 'And if oxygen starts getting low again the fire will show us. It can be our canary in the coal mine.'

Cassidy nods. It's so nice to be able to see everyone properly again after hours in the darkness.

The ceiling of the cave is a few metres above our heads and a few cracks extend even further up into the rock. The smoke from the fire coils towards the cracks and is drawn out to the surface.

The cave itself is about ten metres squared with an uneven, jagged floor and roughhewn, black stone walls. Aravis and Tumnus are still curled up together where the floor and wall meet. I stumble across to them, having somehow forgotten all about them when the storm hit. I call Aravis's name but she doesn't respond. They lie together, motionless and showing no signs of life.

They brought us to this shelter, somehow able to detect that the storm was coming, and then we forgot to share our oxygen with them. I pick Aravis up and hold her close to me as a deep sob escapes from my chest.

JOAQUIN

Lucio holds Joaquin's hand tightly. 'Are you ready for this?' he asks.

'I think so,' Joaquin replies, squeezing Lucio's hand gently.

'Whatever happens, you're still my number one,' Lucio says, pecking Joaquin on the cheek.

'And you're my number one,' he says. 'But it's time to turn this couple into a trio.'

Lucio nods.

They enter the office and sit opposite a fat man in a cheap suit. He looks them over disinterestedly and straightens his tie before speaking.

'You want to adopt a child?' the man asks in a monotone voice.

'Yes,' Joaquin says, a quaver of excitement in his voice.

'Okay, I'll put you on the list,' the man says. He closes his eyes briefly. 'Okay, done.'

'Don't you need to ask us some questions, check our eligibility?' Lucio asks.

'All of that is done if there is a child available for adoption. Currently you are…' he closes his eyes again and Joaquin and Lucio wait patiently. 'Number five-million-four-hundred-thousand-and-seventy-five on the list. We will let you know if you make it to the top.'

'Are you serious?' Joaquin asks incredulously.

'Is there an international registry now or something?' Lucio adds.

'No, that is just here in Peru. *The Blight* continues to affect more and more families,' the fat man responds.

'There are five million couples waiting to adopt a child?' Lucio asks disbelievingly.

The man nods.

'But most people have other options,' Joaquin says indignantly. 'Straight couples can do IVF, use artificial wombs. Gay couples should be given priority.'

'You have other options too,' the man says patiently. 'Have you ever considered fostering a child? That's much less popular.'

'We've talked about it, but thought adoption would be easier,' Joaquins says.

'But if it is less popular, then why not? Add us to that list as well.' Lucio says. 'We can think about it some more while we wait to hear from you.'

Joaquin nods. 'Yes, please put us down for fostering as well.'

The man closes his eyes, inputting their names. 'Done. You are one-thousand-and-thirty-second in line. We'll be in touch.'

'Thank you very much,' Lucio says leaning forward to shake the man's hand.

'I should warn you,' the fat man says, accepting the handshake. 'Foster kids are usually quite a bit older and often come with baggage.'

'Thanks for the heads up. We look forward to your call,' Joaquin says, shaking the man's hand also and turning to leave.

It's a year before they are contacted by the department, but after a few more interviews and screenings, they are allocated twins to care for. Mayra and Azucena are six years old and arrive malnourished and fearful. They have deep bags under their eyes and they watch Joaquin and Lucio cautiously.

'It's okay,' Joaquin says. 'We're here to look after you.'

'You can tell us anything you want,' Lucio says. 'But you also don't have to tell us anything at all.'

The two girls remain silent, exchanging a look.

'Is there anything you would like to ask us?' Lucio asks.

'Who is going to be our new mother?' Azucena asks.

'Well, you're not going to have a mother,' Joaquin replies. 'You will have two fathers. Is that okay?'

The girls don't reply.

'Would you like to see your room?' Joaquin says.

The girls nod.

Mayra and Azucena remain shy, but as Lucio and Joaquin care for them with love and patience, they slowly begin to open up.

'Thank you for all the food that you give us,' Mayra says after dinner.

'Of course,' Lucio says.

'But you don't have to thank us,' Joaquin adds. 'That's just part of being our daughters.'

'It wasn't like that at our old place,' Azucena says.

'They gave us one meal a day. We had to find scraps in bins when we got hungry,' Mayra elaborates.

'I'm so sorry to hear that,' Joaquin says soothingly.

'What we have is yours,' Lucio adds. 'Unless we are all starving, you will have plenty to eat.'

'Thank you,' Azucena and Mayra say together.

Joaquin and Lucio share a look of concern. They'll never fully understand what these girls have been through.

Eleven months later, Joaquin takes the girls to the doctor for their yearly check-up. She is calm and patient with the girls and after some initial hesitation, they warm to her. Both girls have put on significant weight in the last year and the doctor is happy with their overall progress.

'You have looked after them well, but would you mind if I talked to you in private?' the doctor asks with a smile.

'Of course. Are you girls happy playing here?'

They smile at Joaquin and nod happily.

'I noticed on their files that there was some information from before they came into your care,' the doctor says, once they're in the next room. 'Did you receive any information about this?'

'No,' Joaquin replies shaking his head. 'We could see that they were malnourished, but we didn't see any official medical records or anything like that.'

'Well, in that case, I think you had better know that both girls presented with evidence of being sexually assaulted.'

Joaquin is sick to his stomach. How anyone could mistreat two innocent girls like Mayra and Azucena is beyond him.

'Thank you for telling me. I can assure you that my husband and I will continue to care for them the very best that we can.'

She nods. 'While you're here, would you like to have your own health check-up? We've had some huge advances recently. There are a lot of things that we can detect with a simple blood test now.'

Joaquin shrugs. 'Why not?'

Three days later, Joaquin receives a phone call from the doctor while he and the girls are at the park. Azucena waves at him excitedly as she flies down the slide.

'Joaquin, I am sorry that it's taken me so long to get back to you,' the doctor says.

'No worries,' Joaquin says casually.

'I wanted to double check before calling you, because I'm afraid I have some bad news.'

Joaquin takes a deep breath, suddenly dreading what she is about to tell him.

'I am afraid that we found cancer cells in your blood. Initial tests suggest that they may originate from neural tissue, but you will need to come in for more testing. I'm so sorry.'

Joaquin lets the tears run freely down his cheeks. The C word. With all the advances they've made, they still haven't been able to come up with a universal cure. If the doctor's calling, it's one of the bad ones.

'Joaquin. Are you there?' the doctor asks.

'Yes, doctor. Thank you and goodbye. Lucio will be in touch.'

He hangs up the phone and Mayra runs over to him.

'What's the matter, Daddy?' she asks innocently.

'Nothing for you to worry about, my sweet. Now go and fetch your sister. It's time to go home.'

Over the next year Joaquin has test after test and they try different drugs and therapies to limit the growth of the tumours in his brain, but ultimately nothing has a lasting effect.

As the tumours grow, they impact Joaquin's abilities to walk, eat and talk. He is confined to a wheelchair and Lucio must care for him as well as the girls.

'Do you remember what I used to say about cancer,' Lucio asks while he helps Joaquin off the toilet.

'We'd…all…get it…eventually,' Joaquin manages to say.

'Yup. I wonder when I'll get it?' Lucio muses. 'Anyway, would now be a bad time to say, *I told you so*?'

Joaquin laughs, the sound coming out like a wheezing cough.

'You're still my number one,' Lucio says kissing Joaquin gently on the head.

The following day, they have their meeting with human services to ensure that they are still able to care for Mayra and Azucena appropriately.

Lucio assures them that he is coping, while also managing Joaquin's care.

'What is your income at the moment?' the human services representative asks.

'We're living off Joaquin's superannuation,' Lucio replies.

'So, you have no job and a sick husband to care for full time?' they ask sceptically. 'I'm afraid it's not looking good. I'll send through my decision by tomorrow.'

'They...can't,' Joaquin says after they leave.

'They can,' Lucio says angrily. 'And they probably will.'

Human services return the following day to take Mayra and Azucena away.

'You can't!' Mayra screams as she is dragged by the arm towards the front door. 'They're my dads.'

Azucena kicks the man who is trying to shepherd her out the door in the shins. She ducks under his arm and races across the room to Joaquin.

'We love you, Dad,' she says with tears in her eyes. 'Don't let them take us away.'

'Bring her over here,' Lucio says, gesturing to the worker who is man-handling Mayra. He lets go of her begrudgingly and she rushes across to them.

'Listen carefully,' Lucio says. 'We don't want you to leave either, but there is nothing we can do. You will both have to go with these men. I'll visit you.'

Mayra shakes her head, and Lucio pulls her into a big hug.

'Now, say goodbye to Joaquin and go outside,' Lucio says.

They kiss Joaquin tenderly on the cheek and turn away.

'Better,' Joaquin says. 'You…don't…watch…me…die.'

Azucena wipes a tear from the corner of her eye, and the girls rush outside, holding hands.

It takes Joaquin a further three months to die. The house seems quiet and sad without the girls and Joaquin can see that Lucio is slumping into a depression of his own.

'Time…to…go,' Joaquin says as he lies in bed.

'I'm not ready,' Lucio says. 'I don't want to be alone.'

'You…won't…be,' Joaquin says.

Lucio looks at him quizzically.

'No me…equals…yes…girls.' Joaquin smiles, his eyes becoming glassy with tears.

'I don't understand,' Lucio says.

'Promise…me. You…get…Mayra…Azucena…back.' Joaquin takes a deep breath. His body is exhausted and almost ready to give up. 'You…deserve…children. Promise…me.'

'I promise,' Lucio says. 'I know where they are. I will get them back.'

Joaquin nods and closes his eyes.

Lucio leans close and places a delicate kiss on Joaquin's lips.

'You're…my…number…one,' Joaquin says softly.

'You're my number one,' Lucio echoes, a lump in his throat making it hard to speak.

He watches Joaquin's chest as it slowly rises and falls, until the movement stops and Joaquin lays at peace.

'I promise,' he says, kissing Joaquin on the forehead one last time.

SARAH

CHAPTER 15

Tears run down my cheeks as I hold Aravis in my arms.

'I think they're fine,' Newton says from his position nearby, slumped against the wall.

'What do you mean they're fine?' I say angrily. 'We let them suffocate. They're not moving. They're dead.'

My stomach churns and my mind races as grief threatens to engulf me.

'Look closer,' he replies. 'Wait for it… There.' He points to Tumnus and I see his little chest rise and fall once.

I keep watching, hoping to see another breath, but none comes.

'He's not–'

'Buh, buh, buh, buh, buh… there.'

Tumnus breathes again, more than twenty seconds after the first breath.

'I think they went into some sort of hibernation,' Newton continues.

'But there wasn't enough oxygen,' I say.

'Not for us. But for them, at a very slow metabolic rate, there must've still been enough,' he says. 'Or they had enough stored in their blood.'

I stroke Aravis's neck and then try to find a heartbeat. There is a faint pulse and, roughly counting the seconds in my head, I realise her heart is pumping at about four beats per minute.

'How long will they be like this?' I ask.

Newton shrugs. 'How should I know?'

The rain is still torrential outside as we all drift off to sleep. I am exhausted and sleep deeply right through the night, despite the roughness of the ground.

When I wake in the morning, I have a pain in my side where a ridge in the cave floor has been digging in all night. I also have a crick in my neck, and despite the long sleep, still feel weary and have a massive headache

The fire has burned out, but some light is making its way into the cave. I move towards the entrance, the sound of rain increasing as I enter the tunnel. I reach the outermost extent of the cave and see rain falling lightly from a washout of white clouds. The forest floor is scattered with branches and pinecones that have come down in the storm.

I retreat back into the cave where Poh is trying to rekindle the fire. I grab some food and sit down next to her.

'It was really cool how you dove right into the cave yesterday,' I say to her.

'Thanks, it was nothing.'

'No, but it was. Most of us would've hung back and bickered about who was going in. But you just went for it. It was neat. Thanks.'

She smiles up at me. 'Too bad I'm the worst at making fires,' she laughs.

I teach her how to build the fire from small kindling to bigger branches, and soon she has a solid fire burning.

'Why did I find that so hard to learn in training?' she asks.

'Because we had so much to learn. Because it might not have seemed as important when you weren't living and surviving in the reality of a new world?' I guess.

'Yeah, maybe,' she mumbles.

'Plus, it was all pretty rushed. I can't believe we didn't even do the last month.'

'I guess we'll never know why,' Poh muses.

The others get up slowly and help themselves to breakfast. Not that breakfast is any different to lunch or dinner. I think one of the things that I miss most is breakfast food, and it feeling like a distinct and special meal in the day.

We wait out the rest of the rain and start moving soon after lunch. Aravis and Tumnus still haven't stirred when we leave, but their heart rate has doubled. I tuck Aravis into the pouch and carefully place it over my head. The pain in my neck throbs menacingly, so I shift the strap of the pouch to my shoulder instead. It's not comfortable, but it'll have to do.

We work our way back to the shoreline of the lake and continue clockwise on our route.

I almost bump into Cassidy when she stops in her tracks ahead of me.

'Hey, Smithy, there were towns on this lake weren't there?' she asks.

'Of course,' he replies.

'Look at this,' she says bending over and pointing to something sticking out of the ground.

I kneel down beside her and see a small piece of pink plastic buried in the earth. Cassidy grabs the trenching tool and tries to dig it out.

As she pushes the shovel into the ground there is a crunching sound as it hits something buried. She changes angle to dig a bigger section and unearths a crumpled plastic bottle. It is full of dirt and a pink straw, the top of which we saw sticking out of the ground.

‘Fucking plastic,’ Meaghan mutters.

‘But we'd stopped using plastic for things like bottles,’ Poh says.

‘Then this is from before even our time,’ Cassidy says. ‘A relic from the early twenty-first century.’

‘Fucking idiots,’ Meaghan says.

‘Alright, we get the idea, thanks, Meaghan,’ Newton says.

‘But they *were* idiots,’ Meaghan says. ‘It took them so bloody long to stop using plastics.’

‘We weren't much better,’ I say. ‘Sure, we made the R.E.Generation Project work. But our politicians were still bickering about how to address the climate crisis. Businessmen in the most part were still pushing products rather than trying to be part of the solution.’

‘Plus, all our food was packed in single use plastic,’ Newton adds.

‘Humans suck,’ Poh says.

‘That's true,’ Cassidy says sadly.

‘But not us,’ I say. ‘We gave up those lives to make something better. We gave up loads. We have to be better.’

‘Well said,’ Karim says with a smile.

‘Do you think there's more stuff buried?’ Smithy asks.

‘Probably,’ Newton says.

‘Anything useful?’ Smithy adds.

Cassidy holds up the plastic bottle. ‘We could've used this if I hadn't stabbed it.’

The bottle has a large cut where she impaled it with the shovel.

‘We won't know unless we give it a go,’ I say. ‘Look to see if there are any remains of any buildings, and then dig there.’

We divide into three teams and spread out, looking for signs of civilisation. We don't have enough shovels, so Cassidy and I use the axes to cut some long stakes for digging in the ground. Then Smithy and I head towards the pine trees, leaving Cassidy and Newton to search the grass.

We spend a few minutes searching for building remains, but

like at our original camp, the forest has destroyed the buildings that were here with any remains largely buried. We do stumble across some old red bricks, the mortar long since crumbled away to nothing, and decide to start digging.

Smithy can't help much with the excavations, so he watches as I dig the stick into the earth and turn over the soil.

'Hey, Smithy. You know how you lost your arm?' I ask.

'Yeah. It's pretty hard to forget,' he replies.

'I know. I was just thinking, I don't think I've heard you complain about it once.'

'Complaining about it isn't going to bring it back,' he says sadly. 'This is my new reality. It's weird how quickly you get used to it. But I'm also reminded everyday of new things that I can't do anymore, or things I need to learn to do right-handed.'

'Well, I think you're doing an amazing job,' I say with an encouraging smile. 'A lot of people would've given up or focussed on the negatives. But you, you're a fighter. I'm so glad you're here with us.'

'Thanks, Sarah. I'm gonna keep on trying,' he says, unable to keep a broad grin off his face.

'Found something,' I say as my stick hits an obstacle in the soil.

I wiggle the stick under the obstruction and pry it to the surface. Unfortunately, it's just another brick. I keep digging, hoping to find something more useful. It's dirty work, as the overnight rain has made the soil muddy and waterlogged. Smithy takes to scooping the mud that I've dislodged into a large pile, helping me excavate deeper, but getting horrendously grubby in the process.

He also makes a pile of bricks, because for twenty minutes that's all that I unearth. Once we get below the collapsed wall, however, I start to find some random objects.

'What's that?' Smithy asks as I pull a plastic, cylindrical shaped object out of the ground. It has a power cord attached to one end, but the metal prongs have rusted away. I wipe some of the mud away and reveal a narrow neck near one end.

'I think it's a stick blender,' I say shaking my head. 'A bloody stick blender.'

Smithy laughs. 'Well, that'll make whizzing up soups easier.'

'No chance. Look, the metal blades have disappeared.'

He sighs. 'Oh, well. I would love a creamy soup though.'

'Me too. How about you take it to the lake and give it a proper wash. See if we can make anything useful from it.'

He leaves me to continue digging and now the finds keep on coming. I dig up a round plastic disk, also with a power cord, and the plastic shell of a toaster. Once again, the heating elements and all the other metal components have wasted away. The plastic itself is degraded and brittle.

Smithy arrives back with Meaghan, Poh and Karim in tow.

'Have you guys found anything?' I ask.

'Nope. But Smithy said that you might be on to something, so we came to help,' Poh replies.

'Is that the base for a kettle?' Karim asks, pointing at the disk.

'You know, I think you might be right. I was wondering what it was,' I say.

They have the trenching tools, so our excavation speeds up. The next items that we find are a collection of plastic storage containers. Some are broken and unusable, but others look like they'll still be able to hold things for us.

Meaghan and Smithy take them to the lake to wash and sort, as Karim, Poh and I keep digging. We find several other old and decrepit kitchen items like bottles of cleaning fluid, a bag of plastic bags and some plastic cutlery. None of it is usable as it's either already broken or disintegrates as we pull it from the ground.

We decide to give up and head back to the lake to find the others.

'Are any of the containers gonna be useful?' I ask Meaghan and Smithy.

'They would be if any of the lids matched the bottoms,' Meaghan says in frustration.

'Seriously?' I ask.

'Yup, this is the only container with a matching lid,' Smithy says, waving a large rectangular container in the air.

'Bugger me,' Karim says.

'I guess we'd better go look for the rest,' Poh says.

'Maybe in the morning,' I say. 'For now, let's clean up. Smithy, you look like a mud demon.'

The mud has washed off his hand and forearm, but his clothes are covered, and he has managed to smear it across his face and through his hair.

'I hear it's good for you,' he says.

He walks into the lake fully clothed and rinses the mud off as best he can.

'Look out for crocodiles,' Poh calls.

'There are crocodiles?' he asks, shocked.

'There could be,' Poh says with a shrug.

The rest of us wash our hands in the shallows and laugh at Smithy as he tries to get clean as quickly as possible, without thrashing around and attracting unwanted crocodile attention.

I hear footsteps behind us and turn to see Newton and Cassidy carrying a basket between them. It makes a clinking sound as they place it on the ground.

'What did you guys find?' I ask, curious.

'Glass,' Newton says, beaming. 'It was all smashed, presumably from when the building collapsed, but we dug up a whole bunch of shards.'

'Neat,' Karim says, crossing to Cassidy and giving her a peck on the cheek. 'We might be able to make something with it.'

‘You found some containers,’ Newton says, gesturing to the pots on the ground.

‘Yeah, and some other random, plastic stuff,’ I say. ‘It was all pretty crappy though.’

‘I reckon I could use these though,’ Karim says holding up the stick blender, kettle base and toaster.

‘If you say so,’ I say, unconvinced.

We set up camp for the night, planning to dig some more in the morning before moving on. Poh points out that we didn't walk that far today, and that we could go back to the cave for shelter. The rest of us don't fancy the idea.

‘No, thanks,’ Meaghan says. ‘Too many weird memories in that cave.’

‘Yeah, I was definitely hallucinating for a while there,’ Newton adds.

We sit and chat as the sky gets darker. Cassidy is helping Newton and Karim weave thin, springy pine branches into a basket, and together they construct a large carrier. Once they’re finished Cassidy lines it with a blanket and places Aravis and Tumnus inside. They remain unconscious, but Newton informs me their heart rates are slowly increasing. Smithy and Poh are packing up the clean dishes after washing them in the lake, and Meaghan is sharpening the axes. I gaze around the group, smiling to myself. This life could've been complete hell, but somehow, I've managed to find a bunch of people who are kind and caring. A bunch of people who I like spending time with and I’m happy to call my family.

The following morning, we all return to my excavation site. We look for containers that match the lids that we found the previous day, or for lids that match the containers. After about an hour where we only unearth three new lids, and only two that match, we call it quits.

‘Maybe the lids didn’t match, even before the house collapsed,’ Poh suggests.

‘If it’s anything like my mother’s plastics drawer was, then that’d definitely be the case,’ Cassidy says with a wry smile.

Although there may be more we can excavate, we decide to move on and agree to come back once we’re settled.

We continue clockwise around the lake, now heading roughly west. We don’t see any crocodiles, so we decide that it’s probably safe to bathe in the lake when we set up camp in the evening. Even so, we stick close to the shore, keep the splashing to a minimum and stay alert.

We continue to hunt snakes and enjoy kebabs for dinner.

The following day, we stop, as we are now accustomed to, due to a large snake on the shoreline in front of us. It is Newton’s turn to take the kill, so Meaghan hands him the axe and he steps forward poised to strike.

The snake turns to look at him quizzically, unsure what to make of the large, unfamiliar shape coming towards it. Some of the snakes we have encountered have slithered off into the long grass, and some have hissed aggressively, but most have been docile like this one.

Newton swings the axe and there is a dull thud as it embeds in the earth, the snake slithering out of the way at the last second. It lunges towards Newton, seems to change its mind, and races into the grass with haste.

‘Nice one,’ Meaghan says, laughing and stepping forward to clap Newton on the shoulder.

Newton doesn’t respond. He just stands there as still as a statue.

‘Mate, are you alright?’ Meaghan asks, her concern rising.

He mutters something but none of us are able to hear him.

‘You’ll have to repeat that, mate,’ Karim says.

‘It got me!’ Newton yells, bursting with frustration and fear.

'What do you mean?' I ask rushing forward.

'I mean it bit my leg,' he says patronisingly.

'Show me,' I instruct him. 'I didn't think it even got near you.'

'It was so fast,' he whispers. 'So, so fast.' He tries to roll up the bottom of his trousers, but they're fitted too closely, so he has to take them off instead.

I instruct him to lie on the ground and help to remove his trousers so that he doesn't move the affected leg too much. I inspect his calf, and see two small, red dots piercing the skin. Two drops of blood have already coagulated, sealing the wound.

Cassidy steps forward holding a first aid kit.

'Thanks,' I say gratefully. 'Do we have any antivenom?' I ask her.

'Only back at camp with the Alphas,' she replies. 'Declan stockpiled all the rarer drugs in his cabin.'

I take a bandage from the first aid kit. 'Dick.'

She nods.

I wrap the bandage tightly around the wound, before wrapping another bandage from his foot, over the wound and up his leg. I leave his toes exposed so that I can make sure I haven't done the bandage too tightly and that blood is still flowing.

'Alright, you should be fine now,' I say, trying to keep Newton calm.

'It looked like a brown snake to me,' Newton says dejectedly.

'Me too,' Smithy says, unhelpfully.

'Fuck, fuck, fuck. I'm gonna die!' Newton says angrily.

'I need you to stay calm, Newton,' I tell him. 'We need to keep your heart rate down.'

He takes some deep breaths, trying to follow my instructions.

'It could've been anything,' Cassidy says, trying to reassure Newton. 'You're just as likely to be fine as you are to die.' She smiles sweetly.

'So, fifty-fifty, then. Thanks, Cass. That makes me feel so much better,' he says sarcastically.

I take Cassidy to the side to privately discuss what we do next.

'I can see you two scheming over there, you know,' Newton calls. 'If I'm gonna die you should tell me.'

I wave over my shoulder to dismiss him.

'What now?' I ask.

'Well, usually we'd call for an ambulance and make sure he doesn't move, but...' she trails off.

'If he is poisoned, there's nothing we can do about it.' I say sadly. 'So, we're actually just delaying the inevitable by limiting his movement, aren't we?'

'Not necessarily. By keeping the wound immobilized, we might keep the poison in the limb, meaning that he'll just lose a leg, and not die,' she says.

'But we can't just sit pretty. We have no idea when the next storm will hit,' I add.

'True. It's riskier, but I think we put him in the wheelchair and move on. I think it'd be pretty horrible for him, just waiting to die.'

I nod, not liking it but understanding her logic.

'Alright,' I say as we re-join the others. 'You're not going to die,' I lie to Newton. 'Let's get you in the wheelchair and get on the road.'

Newton goes to stand up, but Cassidy quickly stops him by pushing on his shoulder. 'You don't move,' she says calmly.

We unpack the wheelchair and then the six of us carefully lift Newton up and place him in it. He can rest his legs on the triangle projecting out in front of him, with a blanket covering his legs to keep him warm. We load up his lap with some of the baskets, including the dormant lizards. The rest of us have to carry excess items by hand.

Meaghan pushes the wheelchair and we set off, hoping like anything that Newton will be alright.

'Gosh, you're heavy,' Meaghan says jokingly. 'What have you been eating?'

However, in reality, Newton is as thin as a rake. We have all lost weight since waking up, and Newton was already thin to begin with. Hopefully the new range of food, and therefore nutrition, will stop us from wasting away entirely.

Meaghan pushes Newton for most of the morning, declining multiple offers from Karim to take over. Only when she starts to lag behind does she agree to let him take over pushing the wheelchair.

Newton doesn't show any symptoms as the morning progresses, and by lunchtime, I'm starting to become hopeful that he could be fine.

'The venom will still be in his leg,' Cassidy says in a hushed voice, bringing my hopes back down. 'If we got the pressure right, then the venom could be contained in the lymph nodes, while the blood is still able to flow.'

'So, he could still lose the leg,' I say dejectedly.

'Of course, and we still need to be extra careful to make sure the venom doesn't spread,' Cassidy adds.

We continue our walk along the shoreline of the lake. Smithy is confident that we're close to a small river that will lead us towards the old ski fields and the other R.E.Generation unit. The trick is finding the right river. As we walk along the southern end of the lake, many small streams and creeks empty into it. Due to the recent storm, the streams are swollen, but we're able to wade across them without too much difficulty. Carrying Newton across on the wheelchair is a three-person job, and our main concern is keeping all our food stores dry.

After rounding the southernmost tip of the lake, we have to head north, counting the large inlets as we go.

Meaghan is somehow pushing Newton in the wheelchair once again, and when I ask if she'd like to swap out, she declines politely.

Smithy tells us that the river we are looking for should be at the end of the next inlet.

However, as we keep walking, the lake just keeps getting wider.

'That's not what you described, Smithy. What's the plan?' Karim asks.

'Let me think,' he replies, surveying the landscape.

The large land mass in the middle of the lake that Smithy assured us was attached on this southern side appears now to be an island. A large river flows into the lake a few hundred metres further along, and I don't fancy the idea of building another bridge.

'The lake must be fuller than it ever was before,' Smithy says. 'On all the maps I looked at, there was a town here, and as I said, that island was attached. But with all the storms and a constant influx of water, I reckon the old town area is flooded. Which means, that river is still the one we want. It is just a bit bigger than I was expecting.'

'We don't have to cross it, do we?' Poh asks.

'I shouldn't think so,' Smithy replies. 'Just follow the river and we'll find the others.'

'I would say well done, but I won't believe it until I see the shiny pods of our sister unit,' Meaghan says dryly. 'But if this is the river, Smithy, lead on.'

We follow the river, heading south-west for about an hour before stopping for the day. Everyone is in high spirits. The reality of finding other survivors is now tantalizingly close, and Newton continues to look the picture of perfect health.

'Well, thanks for the ride, guys,' he says with a big grin. 'In the end, that was the easiest day of hiking yet.'

'Speak for yourself,' Meaghan says rubbing the small of her back.

'Thanks for pushing the chair so much today, Meaghan,' I say.

'Sure,' she says, walking away to be by herself.

I follow her, wanting her to know how appreciative I am of her hard work.

'No, seriously,' I say catching up to her. 'It was awesome the way you pushed Newton all day. You really took one for the team.'

'Not really,' she replies, glancing around to see if anyone else is listening. 'I kind've had to, didn't I?'

'No, we all could've pushed a bit.'

'But it was my fault that Newton got bitten, so it was my responsibility to push him.'

'No, it wasn't. What are you talking about?' I ask, perplexed.

'I'm the hunter. I should've trained him better. I got complacent because they all seemed so docile.'

'If you honestly think that Newton being bitten was your fault, then you are ten times thicker than you look,' I say. 'He's a zoologist. He knows what he's doing, and we were all complacent. You're no more to blame than Newton is himself.'

'Thanks, Sarah. It's nice to hear that not everyone thinks I'm a screw up,' she says.

'No one thinks that,' I say fiercely.

'I do,' she says, hanging her head. 'All the time.'

'I would tell you not to think that if I knew it was that easy. But it's not, hey?'

She shakes her head.

'Seriously, though, you powered through the k's today pushing that chair,' I tell her. 'You're a real champion.'

She barks a short laugh. 'You reckon it was enough to win the gold medal?'

'I tell you what, the first gold we find, I'll melt it down and make it into a medal for you.'

She smiles broadly and punches my shoulder. 'You sure you're not gay, Sarah?' she asks teasingly. 'Because you know, I'm starting to like you. You're not as basic and boring as I thought you were in training.'

'Is that a compliment or an insult?' I ask.

'I guess it was an insult thousands of years ago, but a compliment now,' she says with a shrug. 'I'd just take it if I were you, though. I don't dole out compliments very often.'

I smile at her and give her a soft punch on the shoulder too. 'Careful. You wouldn't want too many people to see the kind person you've got hidden away in there.'

She continues into the forest to pee and I head back to the others.

'We're gonna try and find a tall tree,' Cassidy says as she and Karim walk past.

'With Newton out of action, we thought we'd check the weather and see where we're going,' Karim adds.

'No worries,' I reply, although I'm sure that is not all they have in mind. 'I'll check on Newton and see what's going on with his leg.'

Smithy and Poh are collecting firewood and starting on dinner, and Newton is watching on helplessly.

'Whatchya doing, champ?' I ask as I approach.

He looks up at me and smiles. 'Not a lot.'

'Let's take a look at your leg, shall we?'

He takes the blanket off his lap, and I kneel beside him to examine the bandaged leg more closely. His toes are a good colour, indicating adequate circulation.

'I'm gonna put a tight tourniquet around your thigh,' I tell him. 'It's gonna be quite uncomfortable, but I'll take it off again in a few minutes, okay?'

He nods his understanding and I set to work. I tie cord around the top of his thigh tightening it with a windlass. I then begin to unwind the bandage from around his leg. I inspect the snake bite closely, but there is no sign of necrotic tissue or disease.

'Does it hurt?' I ask, prodding the wound gently.

'Not really. It's a bit tender when you poke it like that, but it hasn't bothered me all day.'

'What about elsewhere in the leg? Any sore spots?'

He frowns, contemplating the question, and shakes his head. I prod his leg some more, focussing on the areas around the lymph nodes, but he doesn't feel any pain.

'You know what, Newton,' I smile at him. 'I think you're going to be fine.'

'I know, you told me that this morning,' he replies.

'I lied. To be honest, this morning, Cassidy and I thought you were going to die.' I shake my head and chuckle in relief.

'I know. I thought I was too. But I reckon either that snake has evolved to have no venom, or it decided not to inject me. Just a little love bite.'

'Either way, I think you've been very lucky, and we still don't want to take any risks.'

I wrap the bandage tightly around his leg once more and untie the tourniquet to allow blood to flow into his leg once more.

'I think I owe you an apology,' I say, sitting on the ground beside his chair.

'What for?' he asks curious.

'After you told me that you were gay, I've kinda been ignoring you. I didn't do it deliberately, but I haven't talked to you. I'm sorry.'

'I hadn't really noticed,' he says. 'But thanks for apologising.'

'That was really crappy of me. You opened yourself up to me, and then I cut you out. I was upset because I was interested in you, but it's not your fault. I really am sorry.'

'It's okay, Sarah, don't stress it.' He smiles at me kindly and I know that he's genuine.

'It's stupid, though. That it took you nearly dying for me to realise what a bitch I've been. I promise to be a better friend, starting now. Is there anything you need, nay want? I am at your service, Sir.'

'I'd love a hotdog and a big glass of coke. If you could rustle that up, that'd be great,' he says deadpan.

'No worries,' I tell him. 'I'll be back in a jiffy.'

I mix a packet of dehydrated Bolognese with some root flour and water, roll it into a sausage and place it in a pan on the fire.

While it's cooking, I grab a round, plastic container, fill it with water and mix in some ingredients to make it brown like cola. Once the sausage is done, I carefully put it between two floppy mushrooms and take it to Newton, along with the drink.

'Here you are, Sir. A hotdog and cola,' I say proudly.

He takes a bite of the sausage and grins. 'Not bad,' he says through a mouthful of meat and mushroom. He goes to take a swig of drink to wash it down.

'I wouldn't,' I say hastily, pushing the container away from his mouth. 'I legit used dirt to get it the right colour. Here.' I hand him a canteen of water instead.

'What service,' he says, once he's finished his mouthful. 'All I've ever wanted was an undrinkable drink.'

'You're more than welcome,' I reply with a grin.

As he finishes his meal, a rustling noise starts emanating from the basket beside his chair. The basket wobbles a bit and Tumnus carefully climbs over the rim and onto the ground. He tastes the air and looks towards the food in Newton's hands. Tumnus climbs up the side of the chair and sits in Newton's lap, staring at the food and glancing to Newton.

'Alright big boy, you can have some. You must be hungry after your long sleep.'

Tumnus lunges forward and takes a large bite out of the sausage. He carries his dinner down to the ground and starts nibbling on it contentedly.

I glance inside the basket and see that Aravis is also awake. Her eyes are open, but she hasn't moved yet.

'Hello, gorgeous girl,' I croon to her. 'You gave me quite the scare.'

I reach my hand forward to stroke her neck and she pushes her head towards my hand, seeking my touch.

I hear a commotion behind me and turn to see Cassidy and Karim hurrying back into camp, their cheeks flushed.

'There's a building thing,' Cassidy says excitedly.

Meaghan, Poh and Smithy gather around as they provide more details.

'It looks like a giant, silver pod,' Karim adds. 'Kinda like a much bigger version of our own pods actually.'

'Could be an alien spaceship,' I suggest conspiratorially.

'I hope not,' Poh says.

'Where is it?' Smithy asks, getting back to more important matters.

'It's a few k's away,' Cassidy says. 'We saw it from the top of a tree. I think we would need to follow the river further and then climb uphill a kilometre or two.'

'We'll check it out in the morning,' I say. 'Nice find, guys.'

Everyone is excited about what we might find. Every other building has been totally wiped out, so it makes little sense that this one should've survived. As the evening progresses and I think about it further, my excitement slowly changes to trepidation. We've survived well in this new world, but this pod building is a large unknown that could mean anything. Nonetheless, we have to check it out and see what it's all about.

The following morning, we follow the river, travelling constantly on a slight incline. Karim and I climb trees regularly to keep track of the silver building. It is to the south of our trajectory following the river and is nestled on top of a small hill. After about an hour, we decide that if we keep following the river, we will pass the building, and would then need to double back on ourselves.

'I guess I'll wait here,' Newton says a bit dejectedly. 'No point pushing me all the way up a hill, just to come back down.'

'I'll stay with you, mate,' Smithy offers. 'After all, my balance isn't what it used to be since I lost my arm. And I'm not too keen on climbing a hill if I don't have to.'

‘I guess I’d better stay and keep an eye on you both,’ Cassidy says. ‘We can’t have you guys caught in a surprise attack with no one to defend you. Plus, I want to look at Newton’s leg again.’

‘Alright, we’ll see you back here later today,’ I say. ‘No idea how long we’ll be. If we’re not back by nightfall, maybe you should start getting worried.’

The rest of us set off, climbing up the hill as it becomes progressively steeper.

‘Do any of you have your period at the moment?’ Karim asks, after a few minutes.

We look at him in confusion.

‘Why would you ask that?’ Poh queries.

‘Because you’re the PMS girls,’ he states with a broad grin. ‘Poh, Meaghan, Sarah,’ he elaborates.

‘You’re such a dweeb,’ Meaghan says, rolling her eyes and pushing ahead.

It’s about another hour’s hike before we reach the building. We have a steep climb the last few hundred metres, and must scramble up the slope, holding onto some smaller trees to stop ourselves falling backwards.

Once we reach the building, we are met with a sheer wall of metal that curves towards us. A few small trees have tried to grow next to the rounded building but are forced away from it by its convex curve.

Looking behind us, we have an impressive view over the flatlands that we’ve been traversing. There are even some sporadic views of the lake, but mostly what we observe is a sea of pines. It’s the first view that Meaghan and Poh have had, and they gaze out over the trees with open mouths.

Turning back to the building, I note the matte finish of the metal, and the similarities with our own pods. It’s possible that it was made with the same, stable metal compound as our chemo-pods, designed to last thousands of years without breaking down.

There are no visible joins or seams in the metal wall. Although it seems unlikely, it appears that the whole building has been made from one large piece of metal.

Once the girls have finished admiring the view, we walk around the building, looking for a way inside. The whole structure seems sealed however, with no windows or vents. The large pod is oval shaped, more than a hundred metres long and fifty metres wide. Once we have nearly walked the entire perimeter, we finally notice some defects in the wall. A fault-line in the metal, making a small square, sits within a larger, door-shaped outline.

We pause to inspect the shapes more closely, but there are no other clues indicating how to get inside. Poh traces the small square with her finger and then pushes gently in the middle of it, but nothing happens. She taps the wall around the square, and suddenly a section of wall lifts up, revealing a small cavity. Within the alcove sits a small glass tube that tapers towards the top, forming a hollow point. Inside the tube sits a very thin needle, with about a centimetre protruding above the opening.

'Well, well, this looks interesting,' Karim quips. 'Should we prick a finger?'

'It seems unlikely that it would need a blood sacrifice to allow entry,' Meaghan says.

'But not impossible,' he replies. 'I'm willing.'

'It could be a booby trap,' Meaghan says as Karim reaches his hand forward.

He stops, his hand floating halfway to the needle. 'Why would it be booby trapped?' he asks.

'Well, assuming this was built after we were put under, and assuming the world was ending, it would've been a tumultuous time. I'm imagining thugs, crime gangs. I wouldn't want anyone getting inside something like this, so I'd booby trap it. That's what I'd do.'

'She makes a good point,' I say.

Karim lets his hand fall back to his side.

'Let's do one more lap, looking extra carefully for another way in,' I say.

'Sure,' Karim says sullenly, and we turn to continue searching.

DEALER

'Hey, Boss, what do you want us to do with this lot?' Mugger asks, lifting a heavy crate off the back of a truck.

Dealer rolls his eyes. How many times does he need to tell them? 'The frozen sperm and eggs can go directly into the deep freeze. The ovaries, uteri and testicles will need to be chemo-preserved, before being placed into individual vacuum packs.'

'Right, yeah,' Mugger mutters.

Mugger is evidence in support of the theory of devolution of the human species. However, if that were true, how would one explain Dealer's own genius? Plus, Dealer muses, people like Mugger are essential for businesses like this. He has been with Dealer the longest of all his associates, going right back to when they were pushing drugs rather than ReproGoods. But now, thanks to *The Blight*, they're the most profitable black-market goods of them all.

Even respectable, high-flying types use the black market now, not that any of them would admit it. But with clientele like that looking for contraband, prices just keep going up and up.

'Great job today, Boss,' Greaser quips. She is slender, with an e-tattoo running from her right eye to her chin. It changes colour and shape along with her mood.

Dealer smiles benevolently. He doesn't trust Greaser at all and he can tell there's an agile mind behind those scheming eyes, just waiting for a chance to cut Dealer down. Luckily, she's very good at her job.

'Thanks, Greaser. You too. The intel was perfect,' he says.

Earlier that day, they intercepted a delivery of ReproGoods headed for their main competitor. Psycho, their leader, is gonna be pissed. By claiming the shipment for themselves, Dealer and his crew have not only boosted their own stocks, they've also put a big dent in Psycho's operations.

As Dealer crosses the warehouse to help Mugger unload the truck, his security system sends an alert directly into Dealer's ear.

Unknown persons entering restricted area. Persons carrying weapons and items of detainment. They are most likely law enforcement.

'Shit,' Dealer mutters. 'They must know that we've got fresh goods.'

Dealer has had regular contact with the police, but he always keeps his stock buried where no one can find it. Not even Greaser knows the exact location or access codes. The police have never been able to unearth any illegal goods, so they've never been able to charge him with anything. But with a truckload of illegal goods parked in the middle of the warehouse, they're toast now.

'Mugger, time to ghost,' Dealer calls. 'We've got company.' He turns to warn Greaser, but she has already vanished.

Mugger sprints for the toilet block and Dealer slinks into his office.

'Chocolate cola, artificial aftershave,' he states.

The filling cabinet glides sideways, revealing a metal flap in the floor. The lid lifts up exposing a large hole in the ground. Dealer climbs inside and starts descending the ladder.

'Burnt buttocks,' he calls, and the lid closes over his head. His optical implants activate automatically and allow him to see clearly in the subterranean environment. He climbs down the ladder until he reaches a labyrinth of underground passageways. Luckily, thanks to an e-tattoo map, he knows his way around like the back of his hand.

Intruders have entered the inner sanctum, the security system reports.

That means that they are in the warehouse and will be swarming over the truck. They will seize the goods inside, which is a damn shame, but at least Dealer's other stores will remain secret and safe.

He hurries through the tunnels, aiming for an exit point outside of the compound. Hopefully, he'll be able to slip the net. Hopefully, Mugger and Greaser can escape too.

Mugger has been detained.

Stupid idiot. He never could remember which cubicle housed the entrance to the secret tunnels.

Dealer sprints through the underground passageways, turning left and right rapidly, but always heading approximately north. He calls out as he runs, deactivating trip wires and booby traps with his voice.

Dealer has left the restricted area.

'Yes, thank you, I know where I am.'

He rounds the next corner and slows to a walk, catching his breath while he looks for the tunnel out.

Greaser has left the restricted area.

'Interesting,' Dealer murmurs.

He crawls through a tunnel that is barely large enough to fit him and reaches a dead end.

'Zebra zippers,' he says, and the roof above him slides open.

He stands up, appraising the alleyway around him. Blue and red flashing lights reflect off the walls, but there are no police in his immediate vicinity.

'Burnt buttocks,' he says to close the escape hatch.

He stays in the shadows as he carefully slinks further away. He hears the secret door opening behind him, and he turns to see Greaser standing up out of the ground. She unfolds like a bear waking from hibernation and ready for the hunt.

Dealer turns to run away from her, but a large mass steps out from behind a rubbish skip, blocking his escape.

'Checkmate,' Greaser declares.

The stocky man steps forward and whips a large cord towards Dealer. The cord wraps itself around him, tightening of its own accord and holding him fast.

Greaser bundles Dealer into the back of a vehicle, which speeds off as soon as the door is closed. The best thing about driverless cars is that they eliminated the need for getaway drivers, and one less person taking a piece of the pie means a bigger slice for everyone else.

The car stops and Greaser's brute hauls Dealer out the vehicle and drags him unceremoniously through an abandoned office building. With the population shrinking, there are countless empty buildings available for use. The brute drops him into a swivel chair, Greaser ties him down firmly and then they both leave the room.

'What now?' Dealer calls after them. 'You can't just leave me here.'

A man chuckles behind him. 'Of course they can.' The voice is sibilant and menacing. He's heard it described before, but never had the chance to hear it himself.

The man slowly turns Dealer's chair around until they are face to face.

'Psycho, the pleasure is mine,' Dealer says.

Psycho sticks out his tongue and slowly retracts it, staring at Dealer intently. His tongue is forked like a snake. Some people say that he was born with it, while others say he sliced it in half himself. He has e-tattoos on his face that shimmer like reptilian skin.

'There are three ways that we can do this, mister Dealer,' Psycho says, ignoring Dealer's platitudes. 'The easy way, the fun way, or the very easy way. We'll do it in that order, and you let me know when you want to stop.'

'I don't know what you are talking about,' Dealer protests.

'I would like you to tell me how to get into your secret stores.'

Dealer finds Psycho's hissing voice highly irritating. 'And why would I do that?' he asks.

'Our operations are being merged, your stock is coming under my control and I need to know how to access it,' Psycho replies with a sickly smile.

'Fuck you, Psycho. You'll never get my ReproGoods.'

'That's not a very nice attitude to have. I'll ask you nicely one more time. The location and the access codes, please.'

Dealer stares at him stonily.

'Alright, if you don't want to do it the easy way, then we get to move on to the fun way.' Psycho's smile gets even broader, a wicked gleam coming into his eyes.

Psycho takes a large swig of liquid from a flask and swishes it around in his mouth. He winks at Dealer and spits onto his chest.

'Gross, but that's not going to make me tell you anything,' Dealer says.

'So young, so rash,' Psycho says.

'I'm older than you,' Dealer replies.

'True. But not smarter.'

Dealer screams in pain as the spit on his top suddenly burns through his clothes and scorches his skin. He can smell cooking flesh and as he looks down his skin appears to be bubbling. The pain is excruciating, and he can think of nothing else as it consumes him.

The pain disappears as quickly as it started, and Dealer realises that Psycho has doused him in an antacid.

Psycho places a bucket on the floor and claps happily. 'The ReproGoods?' he asks.

Dealer shakes his head.

Psycho tuts and produces a large jug of liquid from behind his back. He dips Dealer's hand into the vat and after a few seconds the blinding pain returns. Something about the acid makes everything else disappear. He can't hear, can't see. He can only feel the pain. This time, Psycho lets him suffer for a few minutes before putting Dealer's hand in a different jug which stops the burning.

Dealer's hand is still sore, a numb ache, but the blinding pain has subsided. He looks at his hand and instantly wishes that he hadn't, as he vomits uncontrollably down his front.

He closes his eyes, but the image doesn't go away. All the skin on his hand has been burnt off, and much of the tissue has eroded away. He can see bones peeking through flesh that is charred black and bloody red.

'It's nasty stuff, isn't it,' Psycho hisses. He is enjoying this way too much.

Dealer nods weakly.

'Are you ready to tell me?' Psycho asks.

Dealer shakes his head, his resolve not yet fully broken.

Psycho steps forward holding the jug of acid and motions as if to pour the fluid on Dealer's groin.

'Fuck it, I'll tell you.' Dealer looks at his good right hand. 'There is a map on the back of my hand, the one that you didn't destroy.'

'I'm listening,' Psycho hisses.

'All of the storerooms are marked with an up-arrow symbol, and the code to get in is…' he looks Psycho directly in the eyes. 'Snitches get slaughtered.'

Psycho pours the acid over Dealer's lap and he returns to a world of agony. Fortunately for Dealer, the pain is not localised enough for him to feel his privates as they melt away. Psycho seems to wait an age before applying the antacid.

'No more games, silly man. I'm getting sick of this,' Psycho says. 'What's the access code.'

'I told you, snitches get slaughtered.'

Psycho shakes his head. 'The fun way isn't fun anymore. Time for the very easy way.'

He produces a needle filled with a clear fluid and jams it into Dealer's neck. The drug makes all the pain go away and Dealer gets a warm and fuzzy feeling deep in his gut.

'Thanks, Psycho,' he says beaming.

'Now tell me, please, the access code.'

'Snitches get slaughtered.'

'Fuck me, you were telling the truth?' Psycho asks.

'Of course,' Dealer says, smiling goofily.

With the truth serum coursing through his blood, Dealer tells Psycho all about the booby traps in his underground lair and provides a complete stocktake of his goods. Psycho separates Dealer's good right hand from his arm using an incredibly sharp machete and one well-placed hack. Dealer watches on in fascination, rather than horror, as the drug continues to mask the pain and distort his thinking. Once they are finished, Psycho unties him and walks him to the ground floor.

'Thanks for your help,' Psycho says. 'You're free to go now.'

'Thanks so much,' Dealer says pulling Psycho into a hug.

'Get off me,' Psycho says. 'That drug should be wearing off any minute.'

Now that he mentions it, Dealer is beginning to get some feeling back. His hands and groin throb and he has to move tenderly down the stairs of the office building.

He's confused why Psycho is letting him go, but he's not one to stare a gift horse in the mouth. Psycho has released him into the alley behind the building and Dealer hobbles away, eager to put distance between him and his rival. He leaves a trail of blood on the ground, which drips from his right wrist, but he doesn't pay it any heed.

As he approaches the far end of the alley, a wall of men and women round the corner and start walking towards him.

'All people that you took from,' Psycho calls from the doorway. 'Brothers, fathers, a sister, fingers...'

One of the mob waves a fingerless hand at Dealer. He vaguely remembers removing them to get information from him. The man holds a baseball bat studded with nails in his good hand.

Greaser is also in the crowd, brandishing a katana. With all the technology in the world, these brutes have decided to go old school.

'You even took Greaser's ovaries from her, but you wouldn't remember that because I gave her a new face.'

Dealer stares at her imploringly, hoping to see a hint of mercy in her eyes.

'I'd say that I'm sorry,' Psycho cackles from his vantage point. 'But I'm not. Have at him.'

Dealer folds in on himself, trying to protect his already burnt and broken body, but there's nowhere to hide and the thugs don't hold back.

After the first few blows, someone takes out his knees and he falls to the ground.

Nails impale his liver and kidneys, a sledgehammer shatters his pelvis and a golf club slams into the back of his head.

Pain lances through his body. Dealer knows the end is near, but one thought manages to rise through the pain and humiliation before his mind and body give way.

All this, just for some bloody sperm.

SARAH

CHAPTER 16

As we walk away from the exposed needle, I hear the unmistakable sound of an airlock being broken.

Turning around, I see Karim retrieving his hand from inside the box, a sheepish look on his face.

'Sorry,' he says, shrugging his shoulders. 'I couldn't resist.'

The larger, door-shaped panel shifts slightly into the building and then slides open, revealing an entranceway. The metal walls of the building are about a metre thick and it is completely dark inside.

'Shall we go in?' Karim asks.

'We've come this far, no point stopping now,' I reply.

'After you, big boy,' Meaghan says sardonically.

Karim steps inside the large pod and lights inside turn on sequentially.

I follow him in watchfully. The entire interior is one open space with curved walls, so it feels strangely like standing inside a large egg. The only flat surface is the floor, which is made of the same metal compound as the rest of the building.

Hundreds of pods that resemble our own chemo-pods lay before us. They are not identical, clearly a newer model, but I have no doubt that they have people inside.

'Woah,' Poh says in awe. 'This is huge. How many are there?'

'I just counted,' Meaghan responds. 'The rows are twenty five by twenty five.'

'So, six hundred and twenty five, then,' Karim says, without missing a beat.

'You didn't seriously just do that maths in your head?' I ask incredulously.

'Of course not,' he responds. 'I just have all of the square numbers committed to memory.'

'Of course you do,' I say, shaking my head.

'What? It's not that uncommon to know *all of the squares,* is it?' He looks around, searching for an ally, but Meaghan just shakes her head and Poh shrugs.

'When you say all of the squares, what exactly do you mean?' Poh asks.

'All of the perfect squares of the integers one to fifty. I really didn't think it was that weird,' he replies.

'Maybe not amongst your friends,' Meaghan says. 'Forty seven by forty seven?' she asks abruptly.

'Two thousand two hundred and nine,' he replies. 'Seriously, square numbers are fascinating though. Once you know two consecutive perfect squares you can work out the next, because the difference always goes up by two.'

'I'm not sure we're interested,' I tell him. 'Let's have a look around.'

'But think about it,' he continues, as we walk away from him to investigate the pods more closely. 'The first four square numbers are one, four, nine and sixteen. The difference between each pair is three, five and seven. If you don't know the next answer, you just follow the pattern. Add nine to sixteen and you get twenty five, which is five by five. *Voilà.*'

I know I've stopped paying attention to what he's saying, and Meaghan and Poh appear to be blocking him out as well.

'Yes, yes, very interesting,' I say absent-mindedly. 'Don't open it,' I say to Poh as she reaches towards one of the pods. 'We don't know if we will be able to activate these chemo-pods, and if we can't, we don't want the guys inside exposed to oxygen. No one opens any pod until we know what we're doing.'

She nods, withdrawing her hand. 'Why haven't they reactivated themselves?' she asks.

'Good question. Different parameters maybe,' I suggest.

'They're at a lower elevation than ours. Maybe there's not enough oxygen here when the toxic storms hit,' Meaghan adds.

'Look at this,' Karim says.

He is standing close to the wall and in front of another alcove with a tube and needle inside. The word *Tse* is written on the wall behind the device.

'And look, there's more,' Karim adds, pointing along the wall.

A line of small squares continues away from the one that Karim has opened. I tap on the right-hand side of the next square, and it slides open, revealing an identical needle inside. The word *Marriot* is inscribed on the wall behind the needle.

'Don't tou–' I begin to say, turning back to Karim, but it's already too late.

He withdraws his hand and I see a small droplet of blood slide down the inside of the tube. When it reaches the bottom, the light inside flashes red, but nothing else happens.

'Worth a try,' he mutters with a shrug.

'Hey guys, up here,' Meaghan calls from the other end of the hall.

We move towards her and see that at the far end, there are more pods beyond those we originally counted. They are shaped differently to the human chemo-pods and larger in size.

'What are they?' Karim asks.

'Could be anything,' Meaghan replies. 'More tools and resources, genetic monstrosities, animals…'

There are thirty of the larger pods arranged in a wide triangle. I step up to the control panel of one and try to find the on switch.

'Just wave your hand over the screen,' Poh says, approaching from behind me. 'But good luck. It seems totally scrambled.'

I swipe my hand over the screen and the monitor flickers into life. The screen is a dull grey with black writing. Unfortunately, all it displays is random letters and numbers that I cannot understand, just as Poh had warned.

'It was the same with the other pods,' she says. 'I've got no idea how to work them.'

'Well, shit,' Meghan exclaims. 'All these extra people and we don't know how to wake them. Karim?'

Karim looks at the screen for a minute but then shakes his head. 'This is way beyond me.'

'The main thing is that we know they're here,' I say calmly. 'That's huge. With the right conditions they might wake up, or we can come back and try again later.' I look around at their sullen faces. 'Cheer up, team. According to Smithy, we're close to our sister group, and at least we know how to operate those chemo-pods.'

'If they're anything like ours, half of them will be destroyed anyway,' Meaghan grumbles.

'No need to be a downer, Meaghan,' I say scoldingly. 'This was a recon mission. We've got some good info but let's do some more scouting before we head back to the others.'

We spend half the day examining the building but turn up nothing else useful. We gather as a group, feeling a bit dejected.

'We'd better head back to the others before they get worried,' Karim says.

We all agree and walk away from the building, wishing we knew more about it.

We traipse down the hill in silence.

'Do you guys wanna know something else about square numbers?' Karim asks.

'No,' the three of us chorus together.

Unfortunately, a minute later he decides to tell us anyway.

'Well, if it's easier, instead of multiplying a number by itself, you can multiply the numbers either side and just add one. For example, ten times ten is one hundred. But it's also nine times eleven, ninety nine, plus one.'

'When would that ever be useful?' Meaghan asks, unwittingly getting sucked into the conversation.

'Forty nine squared might be a bit tricky. But fifty times forty eight is a lot easier. Forty eight times one hundred is four thousand eight hundred, halved is two thousand four hundred. Add one and you get two thousand four hundred and one, which is forty nine squared,' he explains.

'Yeah, because that's much simpler,' I say sarcastically.

'Not my fault if you're a mathematics rookie, Sair,' he replies.

I give him a death stare, but he isn't deterred.

'Also, if you multiply the numbers two away from your desired square and add four, it also works. Or three away and add nine.'

'No one cares, Karim,' Poh says, as kindly as she can manage.

'I care,' he mutters, looking disheartened.

'It is very interesting,' Meaghan says politely. 'But maybe just stick to the mental maths rather than trying to explain it.'

We meet up with the others and report on what we found. They are all excited by the prospect of so many new volunteers, but I quell their enthusiasm when I tell them that we don't know how to activate their pods.

We follow the river upstream for three uneventful days, heading slowly back towards more mountains. The river meanders through the forest and we cover about twice as much ground as if we'd moved in a direct line, but Smithy assures us the safest way to stay on track is to stick to the trusted water source. It also brings with it the perks of having a fresh supply of water constantly available, as well as ferns for food.

Unfortunately, as we've left the lake with its surrounding belt of grasses behind, we no longer have any fresh snake meat.

Newton shows no adverse effects from his snake bite, and we're all fed up with giving him a free ride, so he returns to hiking with the rest of us. We all gratefully load the wheelchair with the excess items that we've been lugging around in our hands.

Newton returns to his role as weatherman and reports on our progress through the valley and into the mountains. Once we're in the foothills, Smithy tells us to keep our eyes peeled for any signs of the other R.E.Generation unit.

It's not until the following morning, however, that we find anything.

'Look, a broken wall,' Poh says, crouching down and brushing dirt off some bricks.

We fan out and find more traces of buildings, but no chemo-pods or signs of life.

'Nice one, Smithy,' Meaghan says accusingly. 'Now what?'

'Don't get your knickers in a twist. This is clearly the town near to the unit. It should be just a few hundred metres upstream and then a few hundred metres away from the river.'

We continue upriver, Smithy counting steps as we go. Once he is happy, he instructs us to spread out along the river, leaving thirty metres between each person. Once we're all in position we walk as a wide line uphill and away from the water.

It becomes quite steep and I have to concentrate on maintaining my footing while I scramble up the slope. I am so engrossed in climbing that I forget to scan my surroundings for the chemo-pods of the other R.E.Generation unit. Luckily, after only a few hundred metres I hear a call from my left. I relay the message to those further to my right and head towards the victorious searchers.

I have to scramble across the face of the slope until the ground under my feet suddenly becomes level.

For some reason that I don't understand, there are less trees on the flat area, and the ones that are scattered around are thinner and shorter than elsewhere in the forest.

I join Smithy and Cassidy who are standing together by one of the small trees. It's only once I am closer that I realise they are also standing next to a pod that is almost fully submerged in the ground. Looking around I see several other pods sunk into the earth.

'There's another lot along there,' Cassidy says gesturing ahead. 'And there must be even more nearby.'

We wait for the others to join us before discussing the ongoing plan.

'First things first, we need to find my girlfriend, Tahira,' Smithy says.

'You haven't talked about her since we left,' I say touching his arm gently.

'I didn't want to get my hopes up,' he says. 'To be honest, I didn't let myself believe we'd actually find them.'

'But you did, Smithy, you brought us right to them,' Meaghan says proudly. 'I never doubted you one bit,' she says with a chuckle.

Cassidy looks at her in surprise.

'Okay, so how do we find her?' Karim asks.

'Open a pod and ask around I guess,' Poh suggests.

'No time like the present,' Cassidy says, and we all look down at the pod that we are standing over.

'Wait a minute,' I say. Everyone looks up at me. 'Why haven't their pods woken them up? There's no need to rush. Sorry, Smithy, but we need to think this through.'

'Is there any way to check the atmospheric data from outside the pod?' Newton asks.

'I don't think so,' Karim replies. 'But if we open it up, we should be able to read the interior display screens. And I can pull up a history.'

'It seems safe enough for the seven of us,' Cassidy adds. 'Worst case, we wake up just this one person after we've cracked their pod to check the data.'

They all look to me, allowing me to make the final call. 'Alright, do it,' I say calmly.

Karim digs down one side of the chemo-pod to find the access lever, while Cassidy flips open the control panel. The command screen lights up, indicating that the pod is still connected to its power source.

Soon, we hear the seal of the airlock breaking as air whistles into the pod. Meaghan and Newton help Karim heave the lid open, revealing a woman inside. She has a neat pixie, haircut, and has managed to be chemo-preserved with a slight smile on her face. I wish I'd thought of that. I probably had a horrified *have I made a terrible mistake?* look on my face.

Karim leans over her apparently lifeless body and checks the data on the screens.

'Registering safe conditions currently, however an anoxic event was recorded seven days ago,' he reports.

'Is that when the last storm hit?' Smithy asks.

'Sounds about right,' Meaghan says.

'Looks like all other atmospheric conditions are fine, but the pods are not reactivating because of regular anoxic events, occurring every twelve to fifteen days.'

'Do you know what elevation this unit is at, Smithy?' I ask.

'About seven hundred metres,' he replies.

'Do we think that we can assume our pods activated us, and we managed to survive the first two storms, because there is enough oxygen at higher altitudes?' I ask the group at large.

'That seems a bit counterintuitive, doesn't it?' Poh ponders.

'Not necessarily,' Karim says. 'Carbon dioxide is heavier than oxygen. Each of the storms have blown in from the coast. Assuming they bring with them masses of CO2 that is sitting on the ocean, which displaces the oxygen, then it could make sense.'

'All that really matters is knowing whether we can survive the storms,' I say pointedly. 'At our old camp, we were fine. But by the lake, we couldn't breathe, and this chemo-pod is registering low oxygen corresponding with the storms. The worst thing we could do is wake up this unit, then all die next time there's no oxygen.'

'So why were we safe at our original camp?' Cassidy asks.

I feel like we're going around in circles. 'Our camp was at an elevation of more than nine-hundred metres,' I say. 'It's the biggest difference that I can think of.'

'I say we wake them,' Meaghan says. 'Then we can all get to higher ground before the next storm.'

'I say that we go to higher ground, wait for the next storm, and if it's safe, we can come back and wake them,' I say.

'I'm not sure I can wait any longer, Sarah,' Smithy says.

'I know, mate. I'm sorry. But it's the safest thing to do.'

We talk around in circles for another hour. I push hard to try and convince them to wait, but most of them are simply too excited.

'Should we vote on it?' Cassidy asks.

'If we have to, but I think it's clear what the best option is,' I say, getting annoyed.

'Don't be impertinent, Sarah,' Karim says sternly. 'We're each entitled to our own opinions.'

'Yes, I know. Sorry,' I say grumpily. 'Let's vote.'

I lose the vote comprehensively, with only Cassidy supporting the wait option. The others are all eager to meet new people and get on with building this shiny, new society we have been tasked with. Of course, Smithy is particularly keen to find and wake Tahira.

Cassidy sets the chemo-pod into action and we all wait for the pixie girl to awaken.

It isn't long before the needles retract and leave her breathing on her own, every cell in her body slowly rejuvenating. This situation is normal to me now. Waiting for someone to wake up. I must remind myself that for her, it's a brand-new experience. Her rebirth onto this new earth.

She opens her eyes and startles, no doubt shocked at the sight of seven strangers staring down at her. If everything was as she expected, her pod would still be closed.

'Hey, sorry to frighten you,' Smithy says calmly. 'No need to get up yet, take your time. We were just wondering if you know which pod Tahira Riaz is in?'

He asks so politely, as if he has just knocked on someone's front door, looking for an elusive neighbour.

'Err, Riaz. Yeah, Riaz. That name rings a bell. Woah, freaky,' she says, sitting up and inspecting the forest. 'Riaz, right? Pretty sure she was in Zeta. They're over there,' she says gesturing into the forest.

'Okay, thanks,' Smithy says, hurrying off in the indicated direction without a backward look.

'Sorry about him. He's looking for his girlfriend. What's your name, anyway?' Karim asks.

'Yi-Seo,' she says, accepting Karim's hand and standing up out of the pod. 'What group are you guys from?' she asks, looking a bit puzzled.

'Eta, Epsilon and Gamma,' I say pointing out the appropriate people.

'We're from the other unit,' Newton adds. 'We hiked here to look for more survivors.'

'Is anyone else here awake?' Yi-Seo asks.

'Nope, you're the first one,' Meaghan says.

'I'd better go and help Smithy. Nice to meet you, Yi-Seo,' I say, smiling.

She waves and I head off after Smithy, who has disappeared into the spindly trees. Cassidy and Karim follow close behind me.

'What are your skills?' I hear Poh ask Yi-Seo.

'Pottery and baking,' Yi-Seo replies as we leave them behind. 'Basically, I like putting things in hot ovens.'

We catch up with Smithy and find him standing over another submerged pod.

'Weird how all the pods are buried,' Karim speculates.

'A few more thousand years and they could've been covered completely,' Cassidy says.

'Looks like we arrived just in time,' I agree.

'How do you know which one is Tahira?' Karim asks Smithy.

There are seven submerged pods arranged in a circle, and Tahira could be inside any one. The final pod seems to be floating in the air, wedged between two small trees and lifted clear off the ground. The sides of the chemo-pod are buckled, reminding me of Kyle's pod from our group, and I know the volunteer inside is gone. I hope for Smithy's sake that it isn't Tahira.

'We could just open them all,' Smithy says hopefully. 'We're gonna wake them all anyway, so we just keep opening them until we find Tahira and wake her first.'

'That's not a good idea,' Cassidy says. 'Once the volunteers inside are exposed to oxygen they need to be reanimated as soon as possible.'

'Sorry, mate, but Cassidy's right,' I say soothingly. 'It has to be one at a time. But after we wake the first volunteer, they should be able to tell us which pod she's inside.'

He nods his understanding. 'Well, I guess we start with this one,' he says looking at his feet. 'Who knows, maybe it will be her inside.'

I step up to the pod, flip open the control panel and am relieved it powers on appropriately.

'It should be good to go,' I say, and Smithy reaches for the lever to open the pod.

'That's her,' he says excitedly as the lid swings open.

Inside lies a woman with olive skin and slim features. She wears a headscarf made of the same fabric as the rest of our uniforms.

I turn to look at Smithy, but he is frozen to the spot. 'Are you alright, mate?'

'It's not her,' he says sombrely.

Karim steps up to stand beside him. 'But you said–'

'I thought it was,' Smithy says, cutting him off. 'But it's not Tahira. It's her twin, Zuhra.'

'Nope. You've made that up,' Cassidy says, beginning to laugh. She wipes the smile off her face immediately as Smithy turns to look at her, dead serious.

'But the R.E.Generation recruiters wouldn't have allowed that,' I say, trying to understand. 'If they're twins, they'd have the same DNA, that's a terrible idea for a fledgling society.'

'Not these twins,' Smithy says, walking away from the pod and sitting against a tree.

I instruct Cassidy to start the procedure to wake Zuhra up and walk over to Smithy. I crouch next to him. He looks exhausted.

'We will find her,' I say calmly. 'But what do you mean, *not these twins?*'

'Their parents were the founders of the R.E.Generation Project,' he replies.

I wait expectantly, leaving him time to tell me everything at his own speed.

'They had IVF, but they genetically engineered the babies to receive exact opposite DNA from each parent,' Smithy explains. 'That way they are both able to be a part of the R.E.Generation Project while maintaining genetic diversity.' He rests his head against the tree behind him and closes his eyes. 'And their parents get every single one of their genes into the next generation,' he adds as an afterthought.

It's a bizarre concept, but genetically engineered children were becoming more and more common in the years before we went under.

'What's the big deal, anyway?' I ask. 'We wake what's-her-name, she tells us where her sister is and you'll be reunited with Tahira in no time.'

Smithy looks at me tiredly. 'Her name is Zuhra. And I'm not sure my nerves can handle it, Sarah. Every time I see a distorted pod, I assume the worst. I can't think straight. I'm so excited and terrified. I knew that Tahira was in Theta not Zeta, but it didn't trigger when Yi-Seo said she was over here.' He shakes his head, bemoaning his rashness and lack of judgement.

'We'll find her, Smithy. And when we do, she is going to be so surprised and happy to see you.'

'I know, I know,' he mutters, letting his head fall to his chest.

I pat him on the shoulder and return to the others.

'He alright?' Karim asks nodding towards Smithy.

'He'll be fine,' I say. 'This stuff is nerve-racking for all of us, and I think it's a thousand times worse for him.'

Cassidy is kneeling next to Zuhra's pod, checking her pulse and breathing. 'I can't imagine what he's going through,' she says, standing up. 'Poor thing is on an emotional rollercoaster.'

'As long as there's a happy ending, it will have been worth the ride,' I say.

A few minutes later, Zuhra starts to stir. We sit a few metres back from the pod, waiting for her to take everything in. She sits up, surveying her surroundings. She briefly glances at us as she observes the dark pines, buried chemo-pods and bright sky.

'Oh, Mum and Dad, what you wouldn't have given to see this,' she mutters to herself. 'And thank you, Allah, for delivering me to this place safely.'

She stands up and walks towards us. Smithy rises from his place behind us and walks forward as well.

'Smithy, what are you doing here?' she asks, seeing him for the first time.

'Just thought I'd come to say hi,' he says managing to muster enough energy to make a joke.

She rushes towards him and embraces him in a warm hug. 'I didn't expect to see you. You were sent off to the other unit. Seriously, what are you doing here?'

'Our unit was hit pretty bad by the passage of time,' I say. She turns to look at me, regarding me carefully as I continue. 'Only thirteen of our pods were salvageable. Smithy led us here in the hope of finding your unit and boosting our numbers.'

'And here you are,' she says.

I find it hard to get a good read on her. She is warm towards Smithy, but seems a bit cool towards Karim, Cassidy and me.

'Smithy is keen to find Tahira. Can you show us where her pod is?' Cassidy asks.

Zuhra leads us towards the Theta pods. As we approach, the ground becomes boggy and by the time we arrive at the pods, we are all up to our ankles in mud.

'Gross,' Karim mutters.

'Get over it, princess,' I tease, secretly hating the feeling of muddy water swishing around inside my shoes.

'This is Tahira's pod,' Zuhra says, pointing to the closest one.

Cassidy bends over to access the control panel, trying to avoid kneeling in the muddy soup, and carefully keeping her long hair off the ground.

'It's not working,' she says after a few seconds.

I lean down next to her, peering at the control screen. A few lines of pixels have lit up across the panel, but the usual dialogue boxes and interface are missing.

Cassidy presses a button and the pixels change, but nothing useful happens.

'What do you mean it's not working?' Zuhra asks.

'It's fried,' I say. 'The system is totally whack.'

'Shit, shit, shit, shit, shit,' Smithy says, starting to pace in the mud.

'Calm down, let's check the others,' Cassidy says. 'Worst case scenario, we can do manual reactivation.'

'I wanna see her,' Smithy demands.

'No can do,' I say sternly. 'We're not opening the pod until we have a plan. Hopefully, everything will be fine, but we just need to think everything through.' I smile at Smithy, trying to be supportive.

'You better make this right,' Smithy says tersely. 'Promise me you'll make this right.'

'I'm not sure we can promise that, Smithy,' Cassidy says calmly. 'But we're gonna do everything we can, which starts with being calm and checking on the other pods.'

Smithy turns his back on us and walks into the forest. Although he's acting like a child, I feel immensely for him. The torment of being so close to a reunion with his girlfriend and then having that continuously postponed is pushing him right to the brink.

'Can you go make sure he doesn't do something stupid?' I ask Karim.

'Let me,' Zuhra says, stepping forward and putting a hand on Karim's chest to hold him back. 'I'll make sure that he's fine. See you soon.'

She follows Smithy into the forest, leaving me with Cassidy and Karim.

Cassidy raises her eyebrows at me knowingly.

'I know, right,' I say with a sigh. 'Now let's see what's going on.'

PERDANA

Perdana sits in his dingy apartment in nothing but his underwear, playing the latest game console. The platform is completely immersive and integrated with his neural network. His body stays motionless, but it feels to Perdana like he is moving around the virtual reality world. The graphics are so impressive that he regularly has to remind himself that he's in a game. He has been spending more and more time in this world, to avoid having to face reality.

Getting shot is the weirdest feeling. They've simulated the pain to an exact level of discomfort without causing distress and as Perdana's character dies, he feels breathless and empty. The feeling only ever lasts a few seconds before he respawns, but it's exhilarating and horrendous at the same time. It also reminds him of a time when he was younger and he couldn't breathe, like he was drowning, but he can't remember the details well enough to place the memory.

He feels a jolt of pain in his arm and looks for a wound but can't find one. Another punch lands on his arm and he starts to become aware of a new voice in the back of his head.

'Turn this devil machine off,' it crows.

He initiates shut down and is spat back into the real world, his vision returning in time to see his sister lining up another punch, but not in time to avoid the blow. He's also struck by the heady aroma of body odour mixed with something else unpleasant.

'Oi, stop that,' Perdana complains, rubbing his arm.

'Decided to come back to the land of the living,' Gita says disapprovingly. 'You look disgusting. When was the last time you showered? Are you still seeing your psych?'

'What day is it?' Perdana asks.

'Friday,' she responds bluntly.

'Well, I started playing yesterday, so my last shower would've been Wednesday,' Perdana says proudly.

'Gross. Have you disconnected since you started playing?' Gita asks.

'Nope. It's way too captivating in there.'

'No wonder you smell like piss, bro, I think you've wet yourself.'

Now that she mentions it, the chair does seem a bit damp and he really needs to take a shit.

'Don't be ridiculous, I'm fine,' he says, lying to her. 'Why are you here, anyway?' he asks, changing the subject.

Gita looks at him pitifully. 'To give you a chance to make something of your life. Mum and Dad would be so mad to see you squandering the chances they gave you. Dad died to get us to this country, and what have you done with that opportunity? The organisation I'm working for is looking for volunteers. It's kind of like what you're doing with this VR, a totally immersive experience where you lose complete track of time.'

'I'm listening,' Perdana says, intrigued. 'What's required?'

Perdana has already completed his medical testing and now a psychologist sits opposite him, asking questions and explaining the trial chemo-genic process.

'We've successfully frozen mice and chimps using this technology, but you would be the first person put into stasis. Do you understand?'

Perdana nods.

'Our plan is to freeze you for one week before waking you up and testing to see if there have been any negative physiological or psychological effects. If there haven't been, we will put you under for one month, then three months, then six months, a year, five years and so on.'

'Hang on, so all I have to do is sleep and do a few tests?' Perdana asks.

'That's right.'

'And will I remember anything? Will I get bored?'

The psychologist looks at him with a blank expression. 'Honestly, we won't know until after the first test. But we imagine it will be like having a general anaesthetic.'

'Will I get older?' Perdana asks.

'That's what we want to find out. I'm sorry I don't have more answers for you,' the psychologist says with a shrug.

'It's fine. It's pretty cool. Seems easy enough to me. Are you sure this is the right amount?' he asks pointing to the salary written on a contract on the table in front of him.

The psychologist nods.

'Then I'm in,' Perdana says enthusiastically.

The first time that Perdana wakes up is pretty weird. He thinks that the process hasn't worked properly because no time seems to have passed, but they assure him a full week has come and gone. The scientists complete their assessments and are happy to find no adverse effects.

Waking up after a month of chemo-preservation feels the same as waking from a week. The scientists buzz about Perdana and he can't quite wrap his head around the fact that time has passed without him having any notion of it. The scientists' initial findings indicate that his cells haven't aged while he has been asleep, but further tests will be needed to confirm this.

Waking up after three months, six months, a year and five years, he always feels exactly the same. Perdana couldn't tell you how long he had been asleep for if he hadn't known in advance. To him, it's as if only a week has passed since he first went under, but to everyone else, more than six years have gone by.

'How are you feeling?' the original psychologist asks Perdana. He has developed very slight crow's feet next to his eyes in the last six years, and his hair has begun the transition to grey.

'I feel fine,' Perdana says. 'I just have one question.'

'Yes, what is it?' the psychologist asks.

'That salary you showed me. Have I been earning that every year I've been in preservation?'

'Of course. It's going directly into your bank account and will be waiting for you whenever you want to leave the experiment.'

Perdana smiles. 'So, what's next?'

'We would be keen to put you into preservation for longer periods. Maybe fifteen years, and then fifty if you are willing. But of course, that will mean that your family and friends will age significantly without you.'

Perdana nods thoughtfully for a second. 'I don't think you realise quite how easy this is for me. I literally do nothing and all of a sudden you wake me up again. If you're going to keep paying me, I will keep going under.'

The psychologist smiles at Perdana fondly. 'You've made this very easy for us. Thank you. And the data you have provided is invaluable.'

'I should probably say goodbye to my family though,' Perdana adds as an afterthought. 'Do you mind if I go under tomorrow?'

'Not at all,' the psychologist says with a smile. 'Take all the time you need.'

Perdana organises a dinner to say goodbye to his siblings. The four of them embrace him in a big hug. They introduce him to his new nieces and nephews who have arrived since he has been away.

'This is JoJo, and this is Baron,' Gita says indicating her children to Perdana. JoJo is four and wraps herself around Perdana's leg. Baron is still only three months old and doesn't do much at all. Perdana can't help but think of him as a giant blob with arms and legs.

The mood is subdued during the dinner, but Perdana tries to keep the conversation light-hearted.

'Guys, I'm not dying. I'm just not gonna be here,' he says. 'It's totally painless for me.'

'But not for us,' Gita says. 'You might not feel the passage of time, but we do, and we miss you terribly.'

A tender silence fills the room and all his siblings look at Perdana lovingly.

'We are very proud of you, though,' Gita adds. 'And Mum and Dad would be too.'

The fifteen-year wake up goes smoothly. Perdana can't help but laugh at the psychologist with his grey hair and wrinkles. It all still feels so surreal to him that he almost convinces himself that the psychologist is wearing prosthetics and is just pretending.

He lies down into his pod once more and they attach the gas mask that knocks him out before the chemical is injected.

The next time he wakes, however, Perdana instantly knows that something has changed. Instead of a team of scientists waiting for him expectantly, a single man that he doesn't recognise stands before him. He wears a long tan coat and his skin looks weathered and worn.

The man must be in his sixties and looks like he could be related to Perdana. He has the family nose.

'Hello, Uncle,' the man says.

'Hello. And who are you, sorry?'

'Baron. Your sister's son.'

'The blob,' Perdana mumbles.

'Pardon?' Baron asks.

'It's nothing,' Perdana replies. 'But why are you waking me up? Where are all the scientists?'

'Gone. They ran out of money and had to shut down operations.'

'Why didn't they wake me?' Perdana asks.

'I think they forgot about you. But the technology was so successful they've put heaps of people into chemo-preservation. Hopefully they'll be able to repopulate the earth once it settles back down.'

'Why would they need to repopulate the earth?' Perdana asks.

'It's pretty cooked,' Baron says. 'And with *The Blight*, we're not creating enough new children anyway. They think humans might die out within a few generations.'

'You're not serious?'

'I am,' Baron says shaking his head sadly. 'My wife and I can't have children. That's why I started searching my family tree and found out about you. I wasn't expecting to find you, but here you are.'

'Here I am,' Perdana replies.

Baron turns to walk anyway and Perdana climbs out of the pod to follow him.

'What now?' Perdana asks.

'Live. Die. It's all the same these days,' Baron says forlornly.

'What are you going to do?'

'Go back to my wife. Drink wine together. Watch the smoke in the sky together, die together.'

'Hold on a second,' Perdana says, starting to get angry. 'If everything is so fucked, why did you wake me up? Why did you bring me back?'

'So that you could live and die your own way. If I didn't wake you up, you would've stayed asleep forever. You would've been as good as dead.'

'You little shit,' Perdana says.

Baron fixes him with a scathing stare. 'You may be my Uncle, but I'm forty years your senior, so watch your language, boy. You know nothing. But if you're so ready to go back to sleep, here's a lethal injection. Everyone has them these days, *just in case*. Just remember, it's like chemo-preservation, but permanent.'

Baron hands Perdana a plastic tube, and without another word, disappears through the laboratory doors.

Perdana was running away from the world even before he joined the Project. He leans against the wall and slides down to the floor. There's a whole new world outside, the future. Waiting to be discovered. But all Perdana wants to do is escape.

He quickly reads the instructions on the side of the tube. It's scary how easy they've made it. He presses one end of the tube against his thigh and depresses the button on the other end. He feels a slight pressure as the chemical is forced into his body. He sighs.

It isn't long before the darkness returns, and Perdana falls into his final sleep. He's died so many times before while playing on his console, that it feels normal. It also feels exactly the same as being put into chemo-preservation. Having been through the process so many times before, Perdana feels he is more ready for death than anyone who ever came before.

SARAH

CHAPTER 17

We step away from Tahira and investigate the other seven pods that are languishing in the mud. They all have similar distortions on their display screens.

'You reckon the electrics are fried?' Karim asks.

'Something like that,' I reply. 'They seem to have power, but they're totally busted.'

'Looks like we'll have to try waking them up manually,' Cassidy says. 'If they're fine inside those pods.'

'I was afraid you were gonna say that,' I grumble.

'You weren't that bad,' Cassidy says.

'Yes, I was. You were just so good that if I was in your team, I had a fighting chance. With everyone else, my stats plummeted.'

As part of our training, all medicos were required to run multiple tests to simulate manually reactivating a preserved human. We completed the assessments prior to our group allocations and were randomly matched with others to complete what felt like never ending rounds of simulations. In the end, the faces of the other volunteers I was working with blurred into one. Except Cassidy's. She was memorable because I only lost two simulated patients when I was with her.

'Well we need more medicos if we are gonna have a fighting chance with this lot, so that's got to be the priority from now on,' Cassidy says.

We regroup with the others and share what we've found. Meaghan, Poh, Newton and Yi-Seo have investigated the final group, the Betas, and describe a scene of disorder and destruction. Two pods crushed by a fallen tree and the remaining pods scattered upside down and on their sides. It seems the trees grew larger and thicker in that area, leaving little room for the chemo-pods.

Smithy and Zuhra have also re-joined us and Zuhra eyes us all sceptically.

'You said that there were thirteen salvageable pods in your unit. I count only seven people,' she says. 'Where are the others?'

'We had to split up,' I say simply.

'You *had* to?' she asks quizzically. 'Rule four hundred and three: for the greatest chance of survival, always keep all survivors together. What, may I ask, was so bad that you had to leave?'

'Declan,' I say calmly. I don't like her attitude, and I don't feel like we need to answer to her. We're the ones who have already survived several months in this new world.

'Doctor, dentist, douchebag,' Poh adds. Karim and Meaghan look at her in surprise. She shrugs.

'That's putting it mildly,' Cassidy says. 'I know a few other descriptors I would use.'

'Anyway, we can explain everything later, but to save going over everything multiple times, would you mind if we wake some of the others?' I say curtly.

'Not at all,' Zuhra says with a false smile.

'Would you mind showing us which of the volunteers in each group have medical training?' Cassidy asks. 'We'll wake them up first and then they can help us wake up the rest.'

Zuhra leads Cassidy back to the Zeta pods and Yi-Seo shows

me the Delta pod containing a doctor called Tan. Newton helps me open the pod and I'm surprised to find a plain-looking man inside.

Tan isn't hideous, but unlike most of the R.E.Generation volunteers he isn't classically attractive either. His nose appears to have been broken, his face is marked with old acne scars and he wears an ugly black stud in his right ear. It's actually refreshing to see a person without model-like good looks.

I initiate the process and monitor his vital signs as the chemical and oxygen spread through his body, bringing microscopic cells back to life, one by one. Yi-Seo hovers behind me, anxious to see whether Tan will be fine.

'I have done this a few times before,' I assure her.

'Sure, of course,' she says, but continues to fidget nonetheless.

Eventually, I see Tan flexing his limbs and a few minutes later, once he's finished his exercises, his eyes slowly slide open. He looks my way and his eyes lock onto mine, seeming to stare into my soul until I look away uncomfortably.

'I don't remember you,' he says. 'Why don't I remember you?'

'I'm from the other unit,' I say mustering the courage to look at him once more. He holds my gaze with a steely stare.

'I assume you're medical,' he states. 'Seeing as you appear to have woken me up. I'm Tan. Dentist and Doctor.'

Here we go again.

'Sarah, Nurse and Midwife,' I manage to say politely. 'We woke you up so that you can help wake the others.'

'Is it safe?' he asks, finally looking away from me to scan his surroundings.

'Well, we've survived to get here. So yeah, mostly it's safe.'

'Mostly?' he asks, raising a quizzical eyebrow.

'I'll explain later. I assume you can do the rest of the Deltas?' I ask.

He nods.

'I'll see you later then,' I say, hurrying away.

I head towards the Beta pods, hoping that there are some survivors that I can wake-up. Newton shows me the way and Smithy tags along to keep us company.

'After all, the pods are all topsy-turvy, so you'll need help getting them upright,' Smithy says. He seems to be returning to his normal self, determined to do what he can to eventually wake Tahira up.

'And you're gonna help with only one arm?' I ask him playfully.

'You'd be surprised,' he says with a wink.

We find the pods as Newton described, not one sitting neatly on its base.

Despite being disturbed, most of the pods are still half buried in dirt and pine needles. Smithy rushes off to grab some digging tools and Newton and I struggle to lift the fallen tree off the two crushed chemo-pods. With a few big shoves it crashes to the ground beside them.

'Not pleasant,' Newton says looking at the buckled pods.

'Not at all,' I reply. 'Let's get any gear that's inside and then shift it. We need to access the power station underneath.'

We lift up the end of one of the pods and jam a large branch underneath to keep it off the ground. Remembering how Karim reattached Smithy and Meaghan's pods to the power, I manage to reverse the process to disconnect the broken chemo-pod. We swivel it out of its cradle of earth and roll it to one side. An unnerving rattling sound emanates from inside the pod as it turns over. The only thing visible inside the vacant cradle is an outlet connected to a tiny geothermal power plant buried deep under our feet.

Newton and I attempt to move one of the toppled chemo-pods to the exposed power outlet with little success. The pod has been pushed off its own power station by a large tree, leaving a tangle of wires snaking away from its base and disappearing into the roots of the tree.

I inspect the power supply more carefully and am able to detach the tangled mess, leaving an unbroken socket to attach to the power source.

Smithy returns with the trenching tools, as well as Poh and Meaghan, who chat absentmindedly. They become silent as they reach the Delta pods and take in the sobering sight. Meaghan carries a first aid kit.

'Thought you could use a few more hands,' Smithy says.

'Thanks,' Newton replies glumly.

With help from Meaghan and Poh, we manage to lift the chemo-pod clear off the ground and carry it to the exposed power outlet. We place it down carefully, propping up one end so that I can attach the power to the socket. Then the others lower it to the ground and I open the control screen, relieved to see the regular command prompts displayed correctly.

We open the lid to reveal a red-headed woman.

I inspect her from head to toe and find that her right hand has shifted within its cradle, the needles tearing the flesh of her hand as they had with Smithy's arm.

'We do everything in our power to make sure she keeps that hand,' Smithy says protectively.

Unlike Smithy's pod, the tubing running to the injured hand seems intact. I nod acknowledgement to him.

I return to the control panel, tap through the prompt screens and begin the regeneration process. While we wait for her to wake up, we start righting the other pods and moving them closer to the functional power station and source. In total, there are six pods that appear to be sealed and hopefully have living volunteers inside.

I check on the ginger-haired girl regularly, and she appears to be progressing as expected. At one point, Smithy calls me over to her anxiously.

'Look, her hand is bleeding,' he says pointing.

'That's good, really good,' I say, smiling.

'It is?'

'If it's bleeding, it means that the blood is flowing properly, the chemical has spread. Of course, we do need to stop the bleeding. Grab me a bandage.'

As I wrap her hand, the volunteer's eyes flutter open.

Newton has moved to stand beside me and her eyes drift to him.

'Yes. Rangas unite,' she says excitedly. 'Feck that hurts,' she says turning back to me and her injured hand.

'I'm not surprised. But hopefully it'll be fine. I'm Sarah.'

'Ailish. Thank you for this,' she says holding up her hand. She seems bubbly and full of energy, ready to jump up and get to work. 'Although some of my fingers are a bit numb.' She grimaces.

'The needles have probably damaged the nerve. If you're lucky, the feeling will return,' I say.

Ailish offers her good hand to Newton who helps her stand up out of the pod.

'You know, us red heads have to stick together,' she says in a deliberately audible whisper. 'Brunettes like Sarah want to eradicate us. They can't help it. It's in their genes.' She smiles broadly.

He smiles back.

'What a shit-show,' she says looking around at the pods of her fellow volunteers.

'Can you show us who is medically trained?' Poh asks.

She points at the second chemo-pod that was crushed by the tree and I hang my head. Meaghan makes a sound of frustration and Poh shakes her head.

I work through the afternoon, activating the remaining five pods, the others helping with the grunt work. Unfortunately, not all the volunteers inside the pods were as lucky as Ailish.

A man with long, oily hair called Callum had his legs slashed to pieces as he shifted within the pod over the years. The movement also broke many of the needles and the tubing that is supposed to supply chemical to his legs. No matter how we try and shift the needles and patch the tubes, we are unable to get sufficient chemical into his legs, and they never reawaken.

He seems in reasonably high spirits considering his predicament, but I leave Smithy with the unenviable job of telling Callum we will have to amputate his legs.

One of the other volunteers is in even worse shape, with cuts all over her body. We argue the merits of trying to reawaken her for a few minutes, ultimately deciding that having exposed her to the air we may as well give her a chance of joining us in this new world.

As the chemical is pumped into her body and her blood begins to thin, it seeps through countless cuts across her frame. We begin to wrap some in bandages but soon after, her body begins to convulse and spasm, before she becomes ominously still. The chemo-pod retracts the mechanical fist that pumped her heart for her. Without it, her heart is motionless and she lies at peace.

I hear a sob behind me and turn to see Ailish with tears streaming down her face.

'She was one of the good ones,' she says, all bubbliness evaporated. 'Goodbye, Leyla.'

I hang my head in respect for a few moments before moving on.

The last three chemo-preserved volunteers are well off considering the movement of their pods.

Graeme, Nia and Dart join the team, each reacting slightly different to the awakening and slowly getting used to our new world.

'Weird how it only seems like minutes ago that we were being put under,' Dart says, looking around in amazement. He has short, cropped hair and designer stubble. 'And now it's like this.'

'And even weirder that for us that feels like several months ago, but it all actually happened on the same day,' Newton says.

'Share with us your wisdom, oh wise ones,' Ailish says, slowly recovering from the loss of her friend.

'All in good time,' I say. 'Shall we see what the others are up to?'

We regroup with the other Unit Two volunteers, as well as Cassidy and Karim. Tan has awoken the rest of the Delta volunteers and Cassidy is surrounded by Zuhra and six other Zeta volunteers. Amongst them, she points out Sinclair, the nurse who helped her reawaken the survivors.

I do a quick head count: twenty new volunteers in addition to our original seven. It's all a bit overwhelming, if I'm perfectly honest. We give them a brief overview of our time awake and the key things we've learnt to survive.

'What now?' Tan asks.

The originals look to me.

'Simple, really. We survive. Follow your training and we'll be fine,' I say.

'Who made you in charge anyway?' Zuhra asks.

'We did,' Meaghan says. 'Trust me, she's earned it.'

'She might be your leader, but she's not ours,' Tan says, a scowl across his scarred face. 'You guys aren't even supposed to be here, really.'

Not wanting a repeat of my ongoing battle with Declan I act quickly to contain the situation. 'I've been thinking about this,' I begin. 'There's enough of us to begin a real democracy. I propose that we elect a council to make decisions in the best interest of us all.'

'Sounds like a reasonable idea to me,' Tan says.

'It seems easiest to me if each group elects a leader to represent them,' I say.

'What about the Theta volunteers?' Zuhra asks.

'They can elect their council member once they are awake.'

Everyone nods and we break off into our smaller groups, the

other six originals joining me and sprawling out on the pine needle floor.

'Anyone want to nominate for leader?' I ask, opening the floor up for each of them to make a challenge.

'I always say if you're onto a good thing, don't change it,' Newton says, smiling. 'If we're voting, I'm voting for Sarah.'

'No need to vote, I'd say,' Meaghan adds. 'As long as you're happy being on the council, Sair?'

'If everyone else is happy with that?' I ask, looking around the group of smiling faces.

They nod and Karim pats me on the back.

'Do you think the other groups will have decided as quickly?' Poh asks, tongue firmly in cheek.

'Probably not. I mean, they're new to this and...' Smithy begins before realising that Poh hadn't asked the question with any seriousness. He still seems preoccupied, presumably thinking about Tahira's pod stuck in the bog.

'It's weird, isn't it?' Cassidy says. 'There's so many of them.'

'It's certainly gonna be different,' I agree.

We chat comfortably and reminisce about the happy and challenging times over the last few months. Each of us waking at different times, but ultimately coming together and then succeeding in our most recent quest to bring more human life into this new world. Time to restock, reload and prepare for the challenges ahead.

'You guys decided yet?' Ailish asks, appearing from the trees. 'We're all ready and waiting for you.'

'Sorry,' Newton mutters.

'We're coming now,' I add.

We follow her back to the Delta pods, which have become the unofficial meeting grounds and base camp.

'The leader for the Betas is Graeme,' Ailish says kicking off proceedings. Graeme steps forward and acknowledges everyone with a brief wave. He seems to be a no-nonsense sort of a guy, with brown hair, dark eyes and a quiet smile. To me, he looks exactly like a Graeme should.

'The Deltas took an anonymous vote,' Tan says. 'I was elected with four votes, ahead of Emmett and Amber with two votes each. I humbly accept the responsibility of representing them on the council.'

'And I'm the Zeta leader,' Zuhra says. I see a few of her fellow volunteers looking a bit disgruntled, but none of them speak up against her. 'I propose we have a meeting immediately to discuss plans. For example, how we plan to survive the night.'

'We've camped under the stars for the last two weeks and have been fine,' I respond. 'So, I reckon we'll be alright. But yes, I agree, let's meet.'

Zuhra walks off into the trees and Tan and Graeme start to follow.

'If someone wants to make a fire and start dinner, that'd be fab,' I call over my shoulder as I hustle to catch up.

Zuhra leads us to a secluded spot away from the others and sits down against a tree.

We join her and I begin by filling them in on all the key information from our time awake. Zuhra sits with her arms folded across her chest, however, I do notice her smiling and nodding on occasion as I outline our accomplishments. When I get to explaining the storm in the cave and the difficulty that we had breathing, however, she is not impressed.

'You mean to say that it's not safe, and next time there is a storm, we won't be able to breathe?' Zuhra asks angrily.

'If we get to higher ground first, we should be okay.' I reply tentatively.

'Should?' Tan asks sceptically. 'That doesn't fill me with hope.'

'To be fair, I advocated waiting for the next storm, at higher climbs, before waking you up. However, most of the others were keen to reactivate you guys. I'm sorry if we have any problems, but we just have to pray we were correct.'

'There's nothing we can do about it now,' Graeme says matter-of-factly. 'The more important question is what's the plan with the Theta pods?'

'Manual reactivation,' I state.

'And there's no other alternatives?' Tan asks.

'Not sure. We'll ask the others, but I think we might be left with needing to do it ourselves,' I say.

'Shit. I was hoping to never have to do that,' he says.

'Tell me about it,' I murmur.

Tan smiles.

'So tomorrow we wake the Theta group,' Zuhra says, taking charge. 'Then we climb higher up the mountain before the next storm and find somewhere safe to set up our new village.'

'Gonna be cruddy to leave so much stuff behind with our pods,' Tan says. 'We'll have to make multiple trips to collect everything.'

'Unless we build a machine to lift everything up the mountain for us,' Graeme says simply.

I look at him in surprise. 'You could do that?'

'Don't see why not, we've got everything we need.'

'Sounds good to me. I don't think I asked you what your professions are,' I say.

'Engineer. Mechanical and electrical,' he states plainly.

'And I'm an anthropologist and survivalist,' Zuhra adds, feeling full of her own importance.

We discuss more matters and manage to stay calm and civil. Each working out one another's differences and prejudices, but regularly coming to consensus between the four of us. They all seem so serious and glum, but I guess we were similar when we first woke up. Your only thought is on survival and nothing seems safe or permanent. I have to remind myself that I've settled into this world of pine trees and mushrooms and it will take the new volunteers time to feel comfortable and at home in these new surrounds.

While we chat, Aravis trots out of the trees and pounces onto my lap. She ferrets around trying to get into my pocket, where I have a handful of pine seeds stored.

Zuhra jumps to her feet in shock. 'What is *that*!?'

'My pet lizard,' I say calmly. 'Her name is Aravis.'

I dig the seeds out of my pocket and Aravis gobbles them up greedily.

'She seems pretty cute,' Tan says softly.

'We're gonna have enough difficulty feeding the volunteers, we can't afford food to feed her too,' Zuhra says indignantly.

'She feeds herself,' I say defensively.

'Is that what you call hand feeding her precious seeds,' Zuhra says, gesturing towards the pine seeds in my hand that Aravis has almost finished.

I pocket the remaining seeds petulantly. 'We would be dead without her. She and Tumnus can detect the storms before we can. You'll see.'

Zuhra doesn't seem completely convinced but doesn't push the point. I can only hope that Aravis and Tumnus are able to win her and the other Unit Two volunteers over, in the same way that they did with Meaghan.

We return to the others and I am happy to see my friends intermingled with the volunteers of Unit Two. They appear to have made a conscious effort to break down barriers and avoid any 'us versus them' mentality.

We raise the issue of the Theta volunteers and Karim suggests swapping in the electronics from a working pod. Graeme offers to help and they head off discussing plans.

While each of the new volunteers has access to their own freeze-dried food, the sudden demand on our stores of fresh food sees our supply run out at the first dinner. They have made a large fire and a number of pots sit around the rim, with a mix of dehydrated stews, flavoured with mushrooms and pine seeds bubbling away.

I spend the evening trying to chat to as many of the new volunteers as I can. I'm normally very good at remembering names, but today, they all get muddled together. Mateo becomes Minh, and I can hardly tell the difference between Iris and Joyce. I even call Sylvie Sinclair on one occasion, even though Sinclair is a male nurse with a manly jawline and broad shoulders. We have gifted the wheelchair to Callum and told him that we can wait a day or two to see if his legs come good. In reality, I'm just giving him the opportunity to come to grips with the fact that he's gonna lose his legs before we actually take them away.

Karim and Graeme return and shake their heads sadly.

'The system is just too complex,' Karim says.

As everyone pulls out their blankets and tries to find a comfortable pile of pine needles to sleep on, Cassidy pulls me to one side and huddles close, conspiratorially.

'You know what I've missed these last few weeks hiking with the others?' she asks in a hushed voice.

I shake my head.

'Getting to chat with you alone, inside our hut each night.'

'Me too,' I reply, smiling. 'It's been nice to get to know all the others, but I've missed some quiet time.'

'You know what I've really wanted to talk to you about?' she asks.

'Not particularly,' I shrug.

'Karim,' she says, a little nervously.

'We're not teenage girls anymore, Cass,' I reply.

'I know, we're fully grown women who can talk about sex openly and honestly, and boy is he good at it. Better than any of my previous partners, if I'm honest.'

'I really don't want to know. Sorry, Cass. Remember when I told you that Karim is kinda like a brother to me?' I say raising my eyebrows quizzically. 'You'll have to talk with Poh if you wanna share all the grizzly details.'

'Fair enough,' she says. 'But aren't you excited? It'll be your turn soon. It's been a while,' she says with a wink.

'To be honest, I haven't thought about it. Not since Newton told me he was gay.'

'Yeah, tough luck on that one.' She smiles sympathetically. 'So, who do you have your eye on out of the new bunch?' she asks.

'I definitely haven't thought that far ahead.'

'I can see you and Sinclair together. Or if you don't want to be with another nurse, Dart is a looker.'

'You're incorrigible,' I say, slightly amused. 'Let's leave the matchmaking for another day.'

She smiles and we roll over, finding comfort amongst the familiar pine needles.

CARLEE

'Mum, do you think we should be worried about this bushfire?' Kayne asks, looking up from his tablet.

Her son is an inquisitive boy and at thirteen, he is already far more intelligent and studious than Carlee ever was as a child. She rolls her eyes but tries her best to engage with him in the discussion.

'Nah, mate. It'd have to get through hundreds of houses to reach us. And the fire fighters would never let that happen.'

'They say that tomorrow is going to be another catastrophic bushfire day,' Kayne adds.

'I wouldn't worry about it too much, Kayne. It sucks for everyone who has lost their houses, but we'll be alright.'

Later that night, when they're driving home from basketball practice Kayne turns the radio on. The usual program has been interrupted for a bushfire update.

'Do we have to?' Carlee complains. 'I just want some good news.'

'Shh,' Kayne says, listening intently.

This is an emergency warning for: Bullengarook, Gisborne, and New Gisborne. There is a large bushfire in the Lerderderg State Park that is not yet under control. The fire is heading in an easterly direction.

You are in imminent danger and need to act immediately. Enact your bushfire plan now. The safest option is to leave now. A relief centre has been established in Melton at the Melton Community Hall.

Furthermore, there is a watch and act warning for: Macedon, Gisborne South, Riddells Creek, Bolinda, Monegeetta, Sunbury, Clarkefield and Wildwood. This alert is in relation to the same fire. Authorities advise that you stay alert and monitor the situation.

For more information, visit the Vic Emergency website or call on 1800 226 226.

'Turn that racket off would you,' Carlee grumbles.

'But they mentioned Sunbury, Mum. What's our bushfire plan?' Kayne asks.

'We don't have one, mate,' she replies. 'We live in the middle of town. We're gonna be totally fine.'

They wake the following morning to a curtain of smoke. Even after the sun should have risen it remains dark outside, as the smoke carpets everything. Carlee doesn't complain when Kayne turns on the radio. The report is sobering as it outlines properties lost on the outskirts of Gisborne. The firefighters managed to protect the majority of the town centre and thankfully there are no casualties reported. The fire has now passed that town and they've turned their attention to protecting properties and lives in Riddells Creek.

Usually on a Saturday, Carlee would take Kayne to footy, but the match is cancelled because of the oppressive smoke. She hangs a load of washing up inside, to avoid smoking the laundry, and tries not to go stir crazy cooped up indoors. Kayne chats with his friends over the internet and keeps the Vic Emergency website open on his phone.

'Mum, they've upgraded Sunbury to Emergency level, should we think about leaving?' he asks.

'You're starting to bug me, my little boy. Trust the firefighters,' she replies.

'Apparently the wind has changed direction and it's driving the fire southeast, directly towards us.'

Carlee glances out the window, a skerrick of anxiety worming its way into her stomach. The smoke obscures most of her view, but she can see the trees in the backyard swaying in the wind. They have the air conditioner on full, but it's proper hot outside.

'Just keep an eye on that website for us, would you? I'm going to bake some cookies.'

Carlee tries to take her mind off the fire, but it's difficult not to think about it. There have been fires nearby in the past and they've never been in any danger. She nods to herself. The fire would have to burn through hundreds of homes to get to theirs, and it wouldn't be a bushfire anymore because they're not in the bush. They're going to be fine. Totally fine.

'Mum, they've raised the warning again. They're saying evacuate now for Sunbury.'

'They can't expect all of Sunbury to evacuate,' she responds. 'That's forty thousand people. Surely that's just for everyone on the northern edge.'

'It doesn't specify, Mum. Lots of my friends are leaving.'

'Are they just?' she asks uneasily. 'Start packing a bag then. These cookies should be done in twenty minutes We'll leave then if we have to.'

'The advice says to leave now. If we wait it may be too late.'

She ignores him.

Kayne packs some things into a bag while Carlee watches her cookies carefully to make sure they don't burn.

Fifteen minutes later, the fire hits them without warning. Driven by the strong winds and oppressive heat, it treats trees and houses alike as fuel. Inside their three-bedroom home, Carlee and Kayne shelter, hoping for the best, but the fire has no sense of mercy or compassion. By the end of the day, it has claimed their lives, along with thousands of others and continues its charge, out of control and ferocious, through Melbourne's suburbs and into the city itself.

SARAH

CHAPTER 18

The following day, Cassidy, Tan, Sinclair and I assemble in the mud surrounding the Theta pods. Before cracking the first one open, we discuss the best team for the manual regeneration process. It takes three people, each with their own special role to focus on.

One person is required to complete the compressions during CPR while another focuses on the breaths. The final team member uses a hand pump to push chemical through the tubes and needles into the dormant volunteer's body, at a constant pressure and flow rate. It is imperative that all three work perfectly together, otherwise oxygen and the chemical are not dispersed effectively throughout the body and the volunteer risks complications ranging from gangrenous fingers to brain death.

'What were your success rates during training?' Cassidy asks.

'Forty percent,' Tan replies, looking downcast.

'Hey, that's better than me,' I say trying to boost his confidence. 'Mine was only a touch over twenty percent.'

'Yeah, my success rate was about forty two percent,' Sinclair adds. 'And you?' he asks Cassidy.

'Sixty seven, but it's all about team work anyway, and I'm sure we've got the perfect team right here,' she says confidently.

As I performed the worst during training, I get to sit out the initial stages of the first awakening.

We open a pod at random and find an older lady. She is already sporting grey streaks in her hair and wrinkles around her eyes. It's possible that she's in her early forties, but I'd guess she's even older. Each of the volunteers were required to have a fertility check as part of the R.E.Generation Project's medical screening. To have been accepted into the program, I have to assume that this lady has surprising fertility for her age.

Sinclair takes control of the hand pump for the chemical, Cassidy leans over the woman's chest and Tan settles in beside her head. Cassidy instructs the others and they stir into action, silent in their concentration.

After a few minutes, Tan takes over from Cassidy on compressions and I replace Tan, giving Cassidy a well-earned break. It's tiring work, so we rotate regularly, trying to keep a steady rhythm between clinicians. Once all the chemical has been deployed, Sinclair slots into the circle, taking his turn delivering breaths and then compressions.

Ten minutes later, Cassidy instructs Sinclair to stop pumping and we all watch anxiously to see if the woman's heart will keep beating without support. We are relieved to see her chest rising and sinking steadily. We slump to the ground and a few minutes later, get to relax completely when she wakes fully.

Edith introduces herself and we explain the situation to her as it stands. She tells us that she is a nurse and midwife, and I instantly feel a competitive drive, hating myself for it.

We let her sit out the next attempted resuscitation, as she familiarises herself with the future.

We take a few more minutes to regroup and rest before moving to the next chemo-pod. Inside is a young man who looks incredibly youthful and I wonder if he is the youngest volunteer yet. I reckon he'd give Jay some stiff competition for the title.

We repeat the process with the new volunteer, but this time there is no pulse or sign of breathing after twenty minutes. We return to work for a final ten minutes before Cassidy calls us all to a stop.

We spread out on the ground, drained and dispirited. The young man lies motionless in his pod.

The physical exhaustion is magnified by the psychological pressure of literally having these volunteers' lives in our hands.

'What did we do wrong?' Sinclair asks.

'Who knows?' Tan says irritated. 'That's the problem with this. There's no bloody feedback to tell you how you're going.'

The possible complications that can lead to an unsuccessful awakening are many. While the chemical is crucial, it is also toxic and an overdose can cause cell death. An underdose can cause cells to stay dormant and then decay over time. Incorrect CPR can limit the spread of oxygenated blood which is also crucial for cellular re-energisation.

'I think it was my fault this time,' Cassidy says.

'Although we'll never really know,' I say. 'All we can do is try harder and hope for better luck next time.'

I close the lid of the young man's pod, saying a silent apology. His death seems unreal, and I hate how much it doesn't affect me. The reality is that I've faced the mortality of so many volunteers since I woke up that I'm almost numb to it now.

We take a longer break to rest and reset and Edith joins the team, performing CPR with the rest of us.

We continue to have varied success with the remaining pods, waking only three of the next five chemo-preserved humans. I open the last pod and, seeing a female face wrapped in a headscarf, feel a heavy weight in the pit of my stomach.

'Tahira,' I mutter, as much to myself as anyone else.

'Zuhra's sister,' Edith says.

'Yes. And the girlfriend of Smithy, one of our fellow volunteers and friends,' I reply.

Tahira looks both alike and entirely dissimilar to Zuhra at the same time.

We review our processes, ensure that the best team is in position to wake Tahira up successfully and begin.

I am surplus to requirements initially, standing to the side with Tan. After a few seconds, I realise that I'm holding my breath, tense and wishing for success for Smithy.

I let out a long breath and Tan gently grabs my hand, giving it a supportive squeeze.

It's not long before he is required to switch in, but even after he lets go of my hand, I feel his warmth and comfort. We've got this.

I take my turns delivering breaths and compressions. By now, we move like a well-oiled machine, sharing looks of understanding and needing few words.

After fifteen minutes, I start to see worry in the eyes of my fellow volunteers, but I refuse to lose hope. Bobby's heart didn't start until we'd done twenty minutes of CPR.

Five minutes later, the cold fingers of dread creep through my body and after a total of thirty minutes we become motionless as one. No words are needed and I walk away from Tahira's pod feeling numb.

If I was surprised by how little the death of the young boy hadn't affected me, I'm equally shocked by how much Tahira's death bowls me over. I didn't even know the girl. But I do know Smithy. And I know I must be the one to tell him.

I half expected him to be here at the Theta pods all day. Obviously, he couldn't stand the thought of watching.

I move back towards ground zero, feeling like I'm floating across the ground. Words racing through my head, but unable to find the right ones. Unable to plan for the upcoming moment.

Karim greets me, his smile disappearing as he sees the look on my face. He directs me towards Smithy, with a small nod of his head.

I find him sitting under a tree, de-winging pine seeds. He looks up expectantly as I approach. I watch as his face changes expression rapidly. Excitement, trepidation, devastation, defiance, disbelief and then devastation again. Without me saying a word, he's able to understand what's happened.

He stands and starts to walk away from me.

'Smithy, I'm sorry,' I call.

He turns and marches towards me, his only hand clenched into a fist. I stand calmly, expecting a blow but unwilling to stop him.

When he reaches me, he swings his arm, but it becomes limp at the last minute and he lets it drape onto my shoulder. I catch him as he buckles at the knees and I lower him to the ground. He begins to sob uncontrollably and slaps his hand against my chest in despair.

His blows are feeble and not aimed at me specifically. I cradle his head and let him cry until he regains control of his body. He slumps to the side and I kneel next to him, hugging him but unable to offer any meaningful words of support.

'Have you told Zuhra?' he asks, his voice husky.

'Not yet.'

'You'd better tell her,' he says. 'She's the only one who will understand.'

I leave him in the middle of the forest floor and search for Zuhra. Looking back, all I see is a pile of human. A broken man.

Edith has beaten me to Zuhra and when I find them, she has tears streaming down her cheeks. I point Zuhra in the direction of Smithy and find my own place to crash.

Cassidy finds me and wraps me in a blanket and a giant hug.

'It's not your fault,' she says.

'I know,' I reply, without believing it.

'No one is to blame. This really is a harsh new world.'

Aravis and Tumnus join us in the cuddle, as if aware of our feelings and wanting to help cheer us up.

We stay huddled together as the news filters through the camp.

Edith, Bobby, Latoya and Dimitrios, the four Thetas to survive, are welcomed by the other volunteers. The feeling is very different to the previous day, however, with grief the dominant emotion rather than joy.

As we sit and watch the camp, Aravis begins to get restless. I stroke her neck to calm her down, but she jumps out of my lap and starts pacing around, stopping to smell the air.

'Storm coming?' Cassidy asks.

'Surely not, it hasn't been long enough,' I reply.

Soon, Tumnus joins Aravis and they scamper around erratically, but unlike at the lake they don't scurry away.

'What's going on?' Karim asks noticing the lizards.

'No idea. It can't be a storm. I'm pretty sure it's only been nine days,' I say.

'It could be, though,' he says hastily.

'But you said they're every twelve to fifteen days,' I say, recalling what he said when we opened Yi-Seo's pod.

'On average,' he replies, holding his hands up innocently. 'The smallest gap I saw in the computer log was eight days.'

'Aravis and Tumnus aren't running away,' Cassidy says hopefully. 'So maybe it's not a storm.'

'Or maybe there's nowhere safe to run to,' Karim adds sombrely.

A collection of other volunteers, both old and new, have gathered around, drawn in by the lizards' erratic behaviour.

'Alright, everyone, back into your pods,' I say, taking charge. 'Make sure your oxygen tank is accessible and if your pod has power, keep an eye on the atmospheric readings and only reopen your pod once we're back to normal.

The Beta volunteers look to Graeme.

'You heard what she said,' he says plainly, turning to find his own pod.

'What about us?' Poh asks.

'There are spare pods,' I reply calmly. 'Find the one you each fit in best.'

'There's only six spare pods by my count,' Meaghan says.

'That's why I'll be outside with an oxygen tank from one of the buckled pods,' I say, maintaining composure.

'But you'll freeze,' Karim says.

'Maybe,' I reply, smiling confidently. 'But all of you will be safe.'

'Let me do it,' Meaghan says. 'I've got a plan.'

'I couldn't–' I begin, but she cuts me off.

'You can and you will,' she says firmly. 'It's far more important that you survive than any of us. Get Smithy into a pod and then find one for yourself.'

'I'll take the single Beta pod available and Poh, you take the Zeta,' Cassidy says. 'Everyone else into the Thetas.'

I scoop Aravis off the ground and put her in her pouch. She struggles at first, but settles, possibly understanding that I have a plan.

I find Smithy where I left him, now embracing Zuhra. They remain eerily silent.

As I approach, Zuhra turns to look at me, venom in her eyes.

'What do you want?' she spits.

'Storm,' I say simply. 'We need to get inside the pods.'

A primal survival instinct takes over Smithy's body and he rises from the ground without needing any further encouragement. He stands tall and his face becomes composed for the first time since he realised Tahira wasn't joining us.

He walks with me in silence and Zuhra breaks away from us to go to her own pod.

The ground starts to become boggy and Smithy starts to falter in his approach. I support him to the pod of the young man who we were unable to revive.

'Where is she?' he asks, barely loud enough for me to hear.

I point to the adjacent pod.

'I want to see her.'

'Are you sure?' I ask gently.

He nods.

Karim and Newton are preparing to get inside their own chemo-pods and look at me worriedly from behind Smithy's back. They have already lifted the lifeless bodies of two Theta volunteers out of the chemo-pods and onto the ground.

Smithy and I approach Tahira's pod and I open the lid slowly. Smithy falls to the ground next to her, covering himself in mud. He strokes her face gently and a fresh round of sobs escape his stricken body.

The wind has started to pick up, but he pays no heed to the pine needles that whip across his face.

'Smithy, we need to get into the pods,' I say anxiously.

'I can't leave her,' he cries.

'She'll be here in the morning,' I reply.

I nod to the boys and they walk over hesitantly. With their help we're able to lift the young boy's body from his pod and coax Smithy to lie down in his place. Karim disconnects the oxygen from its inbuilt compartment and hands it to Smithy before closing the lid.

'I hope he's alright,' Newton says.

'Clearly he's not,' I say. 'But he just has to keep his wits about him long enough to survive the night.'

Newton and Karim help me take Tahira's body from the pod before taking their places inside their newly acquired pods. Karim gives me one last worried look before lying down and closing the lid.

Breathing is starting to get harder, so I rapidly extract my own oxygen tank and lie down in the pod. I double check that Aravis is still in the pouch around my neck and pull the lid closed on top of me. It makes a satisfying thud, followed by a click as I push a lever to ensure that it's airtight.

Some of the monitors buzz into action once the lid of the chemo-pod is closed, but because the circuits are fried, nothing useful is displayed on the screen. Just some flickering, multi-coloured lines and an annoying humming sound that seems to get louder and louder the more I try to ignore it.

I open the valve of the oxygen tank for a few seconds, allowing it into the entire pod so that Aravis and I are both able to breath.

The next few hours are horrible and pass painstakingly slowly. I am unable to see what is going on outside and even the sounds from the storm do not penetrate the steadfast shell of the pod.

I routinely release more oxygen into the pod and have no breathing difficulties. I fall asleep on a few occasions, resting fitfully and unaware of how much time has passed when I wake up. I'm scared that I will fall asleep for too long, run out of oxygen and never wake up again.

Aravis stays largely still on my belly, shifting her weight occasionally, but seems content to be safe and dry.

The minutes and hopefully hours tick by, and I start to wonder about opening my pod. But how will I know when it's safe? I could risk opening it now and checking. I would only need to hold my breath long enough to peek out, but I would lose all the oxygen currently inside the chemo-pod and I don't know how much I have left in the tank.

I fall into a deeper sleep and wake up to Aravis licking my nose. The air feels thin and my breathing laboured, so I quickly release more gas.

Aravis returns to my belly and snuggles down.

'Thanks, little munchkin,' I mumble.

Later, when I have once again dozed off, I am woken by a banging sound. At first, I think it could be a tree that's fallen on my pod, but it comes again in a rhythmic manner.

I take a large lungful of air, just in case, and pull the lever to open my pod. The lid swings open and I'm surprised to see blue sky above me. Tan's head leans into my view.

'Hello, you,' he says, smiling awkwardly.

'Hey,' I reply sitting up.

Karim, Newton, Poh and the Theta volunteers are being helped from their pods. The ground is scattered with branches that have come down in the storm. The bodies of the un-awoken volunteers that we cast aside the previous night lie on the ground saturated. It's a hideous sight.

'We were wondering if you guys were ever gonna get up, but then someone twigged that you guys don't really have power,' he says.

'Yeah, it was pretty eerie in there,' I tell him. 'Thanks for getting us.'

'You're welcome.'

The ground is spongy from the torrential rain. I look around for Meaghan but can't see her anywhere. Emmett is banging on Smithy's pod, but it doesn't show any signs of opening. Emmett is a weed of a man. He's tall, but he is almost always slouched in a pointless attempt to appear more average.

I stride towards Smithy's chemo-pod. Stupid, silly boy. If he hasn't kept himself alive in there, I'll bloody kill him.

I need Smithy. The others might think that I'm strong, but it's only because of them. If I've lost either Smithy or Meaghan, I'm not sure how I'll cope. I'm not sure I'll be able to go on at all.

LISA-MARIE

The moment Lisa-Marie wakes up, she knows that something is wrong. That something terrible has happened. Not because of a supernatural feeling, but because that's how her chemo-pod was programmed.

More accurately, she knows that the earth has died. Or at least, humanity has. The atmosphere and conditions must have become incompatible with mammalian life. But now the atmosphere is safe to breathe again. The toxicity has disappeared, and earth has been reborn.

Lisa-Marie checks the atmospheric data displayed on a monitor just to her left and gasps. Warnings flash, indicating dangerous levels of carbon monoxide and nitrogen dioxide, and oxygen levels are worryingly low.

Fuck, she thinks. The earth hasn't been reborn and her pod shouldn't have reactivated.

A calm, recorded voice fills the pod. 'Welcome, Lisa-Marie. The R.E.Generation Project wishes you all the best for your new life.'

She checks the digital clock, which reads 22:14, 13/06/3,878. The air temperature is thirty seven degrees Celsius.

Her pod was designed to monitor conditions and only reactivate after a period of toxicity, followed by habitable conditions. The world outside, however, does not seem in any way habitable.

Lisa-Marie sighs and decides that now she is awake, she may as well investigate what's outside.

Remembering her training, she slowly re-energises her body. After completing her stretches, she taps through a menu on one of the monitors. She finds what she is looking for and the oxygen mask that is sitting loosely over her mouth suctions onto her face, covering her airways. The tank releases oxygen at a steady rate, allowing her to breathe freely.

Lisa-Marie grabs the handle next to her right arm and pulls it down firmly. There is a whistling of gas escaping as the airlock is broken. She shoves against the lid of the chemo-pod, and it swings open to reveal the outside world.

She is surprised to see most of the clinical laboratory still standing. There are holes in the roof, revealing a starry night sky, but the walls have held solid. The moon provides just enough light to see and, looking around the room, she sees the pods of her fellow volunteers.

She stands up out of her pod and disconnects the oxygen tank from the chemo-pod. She tucks it under her left arm and explores the room. Her right-hand, trailing behind her, traces along the surfaces of the other pods. Everything is covered in a thick layer of dust and mosses have taken root where water has leaked in through the roof.

None of the other pods show any signs of activation. Although it means she will be alone, she is thankful. She wouldn't wish waking up to this nightmare on any of them.

She explores the rest of the building, finding all of the other thirty one pods in stasis and good condition. She ventures outside and stops in her tracks, the view in front of her incomprehensible.

One of the few R.E.Generation units worldwide to be placed

near a city, she stares out over Salzburg, now dead and lifeless. Bathed in starlight, she sees the shapes of buildings, but no lights shine from within. There are no cars on the roads, no headlights winding down narrow streets. No people heading home after working late at the office. No lovers, praying that tonight is the night they conceive their miracle child.

Lisa-Marie's feet carry her down the small mountain overlooking the city and walk with a mind of their own. She doesn't know what she's doing, or what to do next, she just keeps walking, deeper into the city.

The tarmac and concrete have disappeared, reclaimed by nature, and she walks on a soft carpet of grass. Soon, she is dripping with sweat, completely unused to the hot air.

Many of the buildings are covered in vines and creepers. Some of their walls have given out, crumbling into large piles and giving a glimpse of the buildings inside.

Lisa-Marie half expects to find hundreds of skeletons, but the people seem to have left the city before dying. The plants are less mobile, and she spots many dead trees. Standing up or lying down, no new trees take their place, presumably because they couldn't create any viable offspring.

After an hour of walking, Lisa-Marie is in the centre of the city, surrounded by familiar sights. The Mirabell palace still stands in its grandeur, although the gardens have run wild before suffering poorly. Salzburg castle sits high on its hill, looking out over the city. It is eerie walking the familiar streets on her own. They used to be bustling with Austrians and tourists alike. Lisa-Marie regrets all the times that she wished the tourists would go away.

Most of the bridges across the river have collapsed, but one remains standing. She crosses it and wanders through the old town. Buildings that were already over a thousand years old have survived a thousand more.

She doesn't know what she's looking for, and Lisa-Marie's unasked questions remain unanswered. She is overwhelmed by the state of things, and unable to do anything about it.

Lisa-Marie slowly makes her way back to her chemo-pod. As a Gen2 pod, hers is designed so that it can be used twice. It has enough freezing and un-freezing agent to complete the chemopreservation process again. Once she is back at the laboratory, she re-inserts the oxygen tank and tries to determine what went wrong. Everything seems normal, so with no other options, she reluctantly settles herself into the pod. Making sure that she is positioned carefully inside the mould, she pulls the lid closed. She navigates the internal, control panel and selects *Chemo-preserve.* She lays her hand down carefully and a few seconds later, a gas that definitely isn't oxygen is released into her mask. Moments later, she is completely knocked out.

The moment Lisa-Marie wakes up, she knows that something is wrong. That something terrible has happened. Not because of a supernatural feeling, but because that's how her chemo-pod was programmed.

She checks the digital clock which reads 22:14, 14/06/3878. The air temperature is thirty eight degrees Celsius.

A calm, recorded voice fills the pod. 'Welcome, Lisa Marie. The R.E.Generation Project…'

'Fuck off,' she yells, wanting to cry.

One day later and her pod has woken her again. *Fuck, fuck, fuck!* she thinks. She can't put herself to sleep again. This time, she is awake for good, and she knows exactly what to expect outside.

She retraces her steps from the day before. Only now, instead of feeling eerie, everything feels depressive. A huge loneliness settles over her and she sobs as she walks. Now she knows that she is walking towards an inevitable end, but she's not ready just yet.

At Mirabell palace, she climbs through a broken wall and starts to explore the interior rooms. Everything is covered in dust, but much of the ornate decoration and furniture has been preserved, particularly in the central rooms that are most protected from nature's hungry maw.

She climbs a marble staircase decorated with small cherubs and pushes open a huge set of double doors near the top. She enters a grand hall, its high ceilings adorned with chandeliers and its walls and floors made of marble.

Lisa-Marie waltzes around the hall with an imaginary partner and curtsies her thanks once the silent song has ended. Breathing is becoming more difficult as the oxygen levels in her tank dwindle. She sits at an enormous oak table and pretends to eat a five-course meal. Afterall, if you can't invade a private palace and pretend to be super fancy after the world has ended, when can you?

She wanders out of the hall and finds a small bathroom nearby. She sinks into a copper bathtub, completely devoid of water, and closes her eyes. For a while she has been able to distract herself from the reality that she must face, but now she lets it sink in.

She has already said goodbye to her family, but now she says goodbye to her fellow volunteers, hoping they have more luck than her and survive to see a safer world.

Lisa-Marie takes a deep breath of oxygen and forces her mask off her face. For a few seconds she can't bring herself to breathe out, but eventually the physiological need overcomes her mental will and she exhales fully. Her next breath floods her lungs with noxious gases. She breathes evenly and begins to feel dizzy and nauseous. She leans over the edge of the bath and vomits on the floor. The symptoms last barely a minute before she is unconscious and a few minutes later, Lisa-Marie is nothing more than a body in a bathtub.

SARAH

CHAPTER 19

I open Smithy's pod using the exterior lever and the lid lifts open slowly. Smithy lies motionless in the pod, but I see his chest rising and falling steadily.

'Time to get up, Smithy,' I say quietly.

'I don't want to,' he responds, without opening his eyes.

I glance over at Tahira's sodden body and then back at Smithy. Maybe it is better if he doesn't see her.

'That's okay, Smithy, I understand. You stay where you are, keep your eyes shut and we'll carry you to camp.'

He doesn't reply so I have to assume he understands. Tan, Karim, Newton and Emmett step forward and carefully lift Smithy from the pod, before carrying him away on their shoulders.

I'm told that the Beta volunteers haven't yet emerged from their chemo-pods, so I stroll through the forest to their location.

I knock on Cassidy's pod, allowing her to open it from the inside. In response to the banging noise, a large pile of pine needles shifts behind Cassidy's pod.

I hear the chemo-pod begin to open and step past it to inspect the pine needles more closely. Something moves within them again and I tense up, preparing to face whatever is inside.

A large creature crawls out and I take a half step backwards. The creature begins to stand, and I break into laughter as I see Meaghan's face peering out at me from a muddy and needle-covered body. She drops the blanket that she has wrapped around her and stumbles forward, shivering. I stop laughing and rush forward to support her.

I help her to the centre of camp, where some of the Delta volunteers have lit a fire. Meaghan is damp, rather than wet and with a new, dry blanket, a hot drink and the fire to warm her she is rehabilitated quickly.

'You're mad,' I say, after some red has returned to her cheeks.

'You're the one who was planning on staying out,' she rebuts. 'You'd be dead.'

'What exactly was your plan, then?' I ask.

'I buried myself with an oxygen tank in one of the holes left by a shifted pod. Covered myself in a blanket and a huge pile of needles which kept out the worst of the rain and avoided the wind.'

'Well, I'm glad you're alright,' I say, smiling broadly. I surprise her by wrapping her in an enormous hug.

All the volunteers have come together and I'm happy to see that everyone survived the storm unscathed.

'Although that was unplanned, I'd like to congratulate you all on your quick reactions and your success in surviving your first storm,' I say, needing to project my voice so that everyone can hear. 'The storms, along with finding enough food and safe drinking water, are the major dangers in this new world, but I know that working together, we'll be able to get through the rest of them.'

I receive a half-hearted clap for my little speech, which is more disheartening than encouraging.

'Speaking of food, what's the plan with the dead volunteers?' Emmett asks.

Seeing his segue as a non-sequitur, I respond only to the

question. 'I think it would be kind if we give them proper burials, seeing as we know there's a while before the next storm.'

'Shouldn't we eat them, though?' Emmett asks.

I glance around the group and am glad to see that Smithy and Zuhra have taken themselves away and are paying no attention to the exchange. I am also glad to not have to contemplate the question in any seriousness.

'We can't,' Cassidy says, before I am able to respond. 'The chemicals that we used to try and regenerate them are toxic and therefore their meat would be poisonous.'

Emmett is not the only volunteer who is disappointed by the answer. I'm revolted at the thought of cannibalism, but if we needed the protein, and it was safe, I couldn't really argue against it.

'If we're desperate for meat, there are a lot of snakes by the large lake,' Meaghan says. 'We can send expeditions to bring back smoked meat. Plus, there are a few snakes in the trees as well, but they're much harder to find.'

'If our stores last long enough, we should be able to get some crops growing,' Ailish adds.

'And what do you plan on growing?' Karim asks.

'I'd suggest we start with wheat, some legumes and vegetables. The biggest danger to the crops will be the storms.'

'Wait, what?' I ask, completely confused. 'How can you grow all of that?'

'Each botanist was given a miniature seed vault that they can use to propagate old plants, focussing on food sources,' Ailish explains.

'Ayesha never mentioned anything about that,' Newton mutters.

'That little minx,' Poh adds.

'Don't blame her without the full story,' I say, defending her. 'I bet you can trace it back to Declan.'

'True,' Cassidy adds.

'How many botanists do we have?' I ask.

Ailish, Chris and Joyce raise their hands, or maybe it's Iris. Either way, three people raise their hands.

'Fabulous. Once we've moved camp, we will clear an area of ground so that you can plant your seeds.'

'Don't forget, all the insects were wiped out,' Newton adds. 'You'll have to pollinate everything by hand.'

'Talking about moving,' Tan says, taking charge. I feel like he is somewhat unhappy that I have been driving the conversation. 'The council decided last night that we will move uphill, towards the ski fields. Initially we will take limited supplies, as the safest place for spare items is inside the weatherproof pods. Once we have built some houses, we will bring everything else to our fledgling town.'

I see several smiles and nods as the volunteers set themselves for the task ahead.

We spend the rest of the day burying the dead and preparing to leave the following morning. A sombre mood hangs over everyone and conversations remain muted into the evening.

The move to a safe location proceeds smoothly as we hike up the face of the mountain in as straight a line as we can manage. We mark the trees along the route to ensure we're able to find our way back.

The way is steep at times and especially challenging for the volunteers who are pushing Callum in the wheelchair.

Once we reach a flattish ridgeline, Karim tells me that we've only walked about two kilometres, although it's taken us more than thirty minutes, indicating how slowly we've been moving.

I'd estimate that we've risen at least three hundred metres, which should mean that we're further above sea level than our original pods and camp.

The council is happy with the location and we all dump our packs on the ground, marking the place as our new, chosen home.

The builders, carpenters and stonemasons begin construction work, ably supported by the rest of the volunteers. They decide to build wooden houses to ensure that we all have shelter before the next storm. The mountain is made of a workable stone, however, and the plan in the long run is to upgrade to stone dwellings.

The medical team plans and executes the risky operation of removing Callum's legs. Having had experience with Smithy's arm, Cassidy takes lead, supported by Tan. By making several small cuts, they are able to find the line between dead limb and healthy thigh. By leaving a thin line of dead tissue, they can cut away the dead limbs which have no blood flow or feeling. The risk being that the remaining dead tissue could turn gangrenous.

The operation is painstaking and slow, and far from perfect. The edge of the dead tissue is unpredictable and on a number of occasions, they cut into living tissue, greeted by a scream of agony from Callum. They keep blood loss to a minimum and after hours of intense concentration, have succeeded in removing both of Callum's legs. He has been awake the whole time but has staunchly avoided looking at the operation.

When Cassidy and Tan collapse onto their haunches in exhaustion Callum finally looks down and a steady stream of silent tears run down his cheeks.

The following day, Bobby, the hunter and tanner from Theta that we had to manually reawaken, shows us his fingers and toes. They have started to turn black and the nails are yellow. 'I couldn't really feel them when I woke up, but I thought they'd get better,' he tells us. 'But they've gotten worse.'

We operate to remove the gangrenous digits before the infection can spread. We are able to use some local anaesthetic from our medical supplies and only have to remove the final phalanx of each finger.

Over the next six months, we build a town and society, without much infighting and no catastrophes. We manage to build some shelters before the next storm hits. We are all anxious when it arrives, but thankfully, it is no worse than at our original camp. We keep building and soon we have a large clearing dotted with buildings and an area of farmland that the horticulturists have planted. They tell us that the soil is not the most nutritious and that we shouldn't expect much, but as the months progress, they produce more and more fresh grains and vegetables, with every plant painstakingly pollinated by hand. After the struggles we've had to feed ourselves, to taste pumpkin, bread made from wheat flour and fresh lentils again is like the first taste of lollies as a child. To protect the crops from the storms, we build solid walls around the farm area to resist the gale force winds. We also weave gigantic sheets from thin pine branches, that can be attached to the wooden walls as a temporary roof to keep off the worst of the rain.

The botanists also plant fruit trees, but we will have to wait much longer before we can enjoy the fruits of that labour.

As everything progresses and our town continues to grow, it doesn't take long for people to start talking about reproduction. We divide into couples and select a handful to attempt to have children, while everyone else abstains. We want to make sure that we don't lose half of our workforce in one big hit. Everyone is happy enough with their partner, and I start to see that there are benefits to most people being attractive when procreation is so vital. Cassidy shows the first symptoms of pregnancy, including a month of morning sickness where every day, like clockwork, she walks into the tree line and vomits under the same tree. After this, however, she experiences heavy bleeding and her symptoms subside, indicating an early termination. It's a tough thing to see her go through, but because we're such an open society, at least she has plenty of people to talk to about it. She was so excited at the thought of becoming a mother, it's cruel to see it taken away from her, for the time being, at least. Despite my protestations, she insists on telling me almost everything about her and Karim's sex life and informs me they haven't wasted any time in trying to get pregnant again.

As for myself, I am paired with Tan, which I'm unsure about. I'm happy to sleep with the most unattractive man of the lot. After all, it's transactional and I'm not particularly attracted to any of the men. However, Tan and I haven't always seen eye to eye and I'm wary of how our union would work.

Despite my reservations, our first time together is exciting and surprisingly less awkward than I expected. I didn't realise how randy I was until I had the opportunity to do something about it, but after months without sex, I'm definitely up for it.

To be fair to Tan, he knows what he's doing, taking his time with foreplay and attending to my needs and desires. However, remembering that he also hasn't had sex for a while, it makes sense that he's finishes far too quickly.

I think I surprise him with my eagerness to try again the following day, and regularly after that. After our first time together, he shows good stamina and I begin to look forward to it, knowing that he's able to make me feel wonderful.

It isn't long before I fall pregnant, and when I tell him, he sits holding my hands for ten minutes, just staring at me with a ridiculous grin plastered across his face.

We send teams to continue the excavation near the lake. They don't find much of use, but everything helps. Meaghan also organises regular hunting expeditions to the grasslands around the lake. She never has any difficulty finding volunteers to accompany her. They build a smokehouse near the shoreline and preserve the snakes that they kill, so they can bring meat back to camp. Nothing compares to the fresh snake meat that we had on our original journey past the lake, but the smoked meat makes a healthy snack and adds flavour to stews and other dishes. Despite losing the end of his fingers, Bobby can also use the snakeskin to make small amounts of leather.

No one reports seeing any other lizards like Tumnus and Aravis, however, we do find a species of smaller lizard living in the forest around our town. Newton and some of the other zoologists keep a few lizards in captivity to learn more about how they reproduce and their life cycle in general.

Aravis and Tumnus continue to grow and after five months, are almost as big as their mother was when we killed her. They develop into highly intelligent animals that appear capable of communicating with each other through a combination of sounds and gestures. Cassidy and I spend a large amount of time with them and try, with limited success, to decipher these movements.

True to my promise, they feed themselves, disappearing into the forest a few times a day, returning satiated. They even help with the foraging effort, sometimes dragging large mushrooms into camp.

Finding enough food and water is an ongoing process. Anyone who is not busy with other jobs spends hours hunting the forest for edible fungi, collecting pine seeds and hiking to the river and back to collect water.

The hike back up the steep mountain carrying as much water as possible certainly keeps us fit, but is exhausting.

To solve this problem, Graeme designs a giant pulley system to take things from the first camp, near the pods at the bottom of the hill, to our new town up the mountain. The engineers, and Nia, the blacksmith, construct a forge and build a motor by disassembling some of the chemo-pods. We clear a two kilometre stretch of turf running from the chemo-pods up the mountain, and our woodworkers carve wooden pulleys at regular intervals to run a rope over. Fortunately, some of the volunteers arrived with lengths of rope and this is supplemented by cord we make from foraged grasses. It takes a while to process the grass and prepare the rope, but with everyone working together we are eventually able to make enough.

Not dissimilar to a ski-lift, the system is completed by a large pulley at the top and another at the bottom that is driven by the motor, which is in turn connected to one of the underground power stations. One of the remaining pods is converted into a large wagon, by gutting the interior and attaching four large wooden wheels. The wagon is attached to the rope and *The Machine* is ready for its first run.

All the volunteers gather to watch as the wagon is filled with canteens of water and other supplies. When the motor is turned on, the rope starts to move and the wagon is pulled slowly up the hill.

The volunteers let out a cheer, watching The Machine lug litres of water up the hill for us. The wagon disappears into the distance, accompanied by a few volunteers. We wait patiently until a section of different coloured rope appears, our indicator that the wagon has reached the top of its run.

Using The Machine, we are able to drag the remaining chemo-pods up to our new village along with all the resources stored inside each pod. It also allows us to take water from the river to our town with relative ease. The main thing that we must remember is to have everyone up in the village when storms are expected.

While the storms bring with them their own challenges, some of our most enjoyable times are spent hunkered down together inside our sturdy buildings. We have no difficulties breathing, confirming our theory about safety at altitude.

Although I make a strong effort to mingle with all the volunteers, I still find that I spend most of my time with the volunteers from my own unit. We developed such a strong bond during our time together, before we had awoken the others, that it just feels comfortable being with them. Without even discussing it, we somehow come to a group decision not to mention the large pod building to the others. I can't imagine waking up even more volunteers at this time.

The only original volunteer that I don't spend time with is Smithy. He is completely broken by the death of Tahira and isolates himself in his grief. Initially, I try and comfort him, but he pushes me away, preferring to find solace in Zuhra, who is also grieving the loss of her twin sister. It's sad to lose my relationship with him, but as the months pass and he starts to learn to cope, it's nice to watch their friendship develop from afar.

As for Zuhra, I begin to like her more and more as time goes by. Although she can be opinionated and bossy, I realise that this isn't without reason, and I learn to rely on her wealth of knowledge. She is caring and compassionate with Smithy and helps pull him through his greatest struggle, even though she has every right to be a mess herself. She is incredibly strong and fiercely determined to see the R.E.Generation project succeed.

My pregnancy continues smoothly. I'm not sure if it's because I have his child growing inside me, or just because he is brilliant, but over time, I realise it's not just lust that I feel for Tan.

He is funny, caring and kind. Smart, athletic and strong. He is everything I could ever want in a father for my children, except for his taste in ear piercings. I especially love the way that he turns to putty in my hands when I run my fingers through his hair and massage his head.

Even after I'm pregnant, we continue to sleep together. I take this as a sign that he reciprocates my feelings, but the one thing we never do is tell each other how we really feel. When we're around everyone else, we act in a professional manner, business as usual. But when we lie down together, his arms wrapped around me protectively, hands resting on my swollen belly, I feel his love for me.

MELINDA

'What fucking good is it being married to a billionaire if that money can't buy happiness?' Melinda's husband Cory yells.

'We've tried everything, Cory,' Mel says, despairing. 'Rounds and rounds of IVF, IUS, surrogates and QUS. The only option left is adoption.'

'And I've told you that's not an option,' Cory replies heatedly. 'Besides, who's gonna put their kid up for adoption these days? *The Blight* is stopping everyone from conceiving.'

Melinda looks out the glass windows of their penthouse apartment. Countless dark blotches stain the city scape where previously prosperous buildings stand in complete darkness. No lights and no life inside.

'Well, what do you want to do?' Melinda asks. 'More IVF?'

'Yes, but let's get a new doctor.'

Melinda and Cory wait outside Dr Gopal's office, holding hands tightly. The LED lighting is bright and clinical. The walls are a grotesque dark green and the floors are polished plastic. Made from recycled bottles, bags and whatnot.

When the assistant calls them, they rise from their chairs, stand to their full height and stroll into Dr Gopal's private suites.

Dr Gopal is a short, older man with dark skin and grey hair. He offers them his hand and they shake it politely.

Cory doesn't like the look of Gopal and Melinda wonders if she could break his hand if she squeezed too tightly.

'You have a very interesting CV,' Dr Gopal says, in silky tones. 'It looks like you've tried *almost* everything.'

His intonation on the word 'almost' catches Melinda's attention. 'Almost?' she asks.

'Have you heard of extra-personal-uterine-delivery?' he asks.

They both shake their heads, intrigued.

'Of course you haven't,' Gopal continues. 'It's only very new, and very, very expensive.'

'That won't be a problem,' Cory says.

'What does it involve?' Melinda asks.

'We take a bone marrow sample and use your stem cells to create an artificial uterus, which we suspend in a special fluid. Using IVF, we implant an embryo into the uterus. We are able to monitor the baby's development more exactly and as a bonus, you don't have to carry it around inside you.'

Cory and Melinda glance at each other. There's no doubt in either if their minds. Cory nods.

'Can we do more than one?' Melinda asks. 'Increase our chances?'

'I'm not sure you heard me. I did say that it is very expensive. It's five hundred thousand dollars per process.'

'And I'm not sure you heard me when I said that won't be a problem,' Cory replies.

'Let's start with six,' Melinda says.

Dr Gopal's eyebrows shoot up in surprise. 'Well then, when are you are you able to start?'

'As soon as possible,' Cory says.

'Today would be ideal,' Melinda adds.

'I'm not sure if we have any spare appointments–' Gopal begins.

Cory slides a white envelope across the table. 'I think you'll be able to find a slot,' he says.

Gopal opens the envelope and his eyebrows rise even further than before. 'Would you look at that, an appointment has actually just become available. Follow me.'

Gopal leads them through a back door and down a dimly lit corridor.

'Has this procedure been approved by the medical board?' Melinda asks.

'Not exactly,' Dr Gopal says, without looking back at them. 'But the beauty is that there is no risk for the patient, so there's nothing to lose by giving it a try.'

Cory and Melinda glance at each other and shrug.

They pass a couple waiting on plastic chairs in the corridor.

'Sorry, guys, unexpected delays,' Gopal says to them. 'I'll be with you in a few minutes.'

Dr Gopal gestures Cory and Melinda into another room and asks Melinda to lie down on the bed.

After a quick procedure using a device that looks like a corkscrew, Gopal has the sample he needs. He patches up the wound on Melinda's hip and shows them back to his consulting room.

'In about a month, you will be able to see the uteri we have created and then we can begin the IVF process.'

'We already have some valid embryos from our previous cycles,' Melinda says. 'We'll have them sent to you.'

He nods and they shake hands once more. Melinda squeezes hard and Gopal winces. He has the good grace to smile and say nothing.

Over the next few months, Melinda and Cory become regulars at Dr Gopal's clinic. They have an entire room set aside for their artificial uteri. Six large tanks filled with a bubbling green-yellow fluid stand in two rows. The rich red vascular organs float suspended. Throbbing arteries provide oxygenated blood from external machines.

Two of the embryos fail to implant and a third dies after a few weeks, but three embryos survive and start to grow. Their development is monitored constantly and Dr Gopal is able to respond to any complications immediately.

Melinda loves sitting in the uterus room, bathed in green light and watching the uteri swell. She spends hours watching over them, like a mother tending her nest of eggs.

The weather has been extreme for a number of years now, but Cory and Melinda have managed to avoid the worst of it in their expensive apartment. In fact, one of their favourite things to do is turn off all the inside lights during a storm and watch the water falling from the sky and crashing against their reinforced glass ceiling.

When their foetuses are three months old, one such storm hits, and they lie in bed looking up at the elements. Lightning flashes but they can't hear the thunder inside their soundproof box. Hail stones as large as golf balls smash into the glass above their heads, but not a single crack appears.

'It's a wild one,' Melinda says with a smile.

'Thor must be unhappy,' Cory adds.

'Can you feel the building swaying?' Melinda asks.

'I think I can,' he replies. 'That's new.'

They fall asleep, warm and safe in their sanctuary.

Melinda is woken the next morning by her phone ringing incessantly.

'Turn that off,' Cory mumbles.

Melinda rejects the call but moments later, it rings again. She rolls over with a sigh and answers the call.

'Hello.'

'I think you should come down to the clinic,' Dr Gopal says down the line.

Usually, Gopal's assistant Cheryl deals with all his phone calls, so the fact he is calling himself is cause for concern.

'We'll be there as soon as we can,' Melinda says.

She rolls out of bed and whips the covers off Cory. He groans.

'Who was it?' he asks.

'Gopal. We have to go.'

They dress quickly and their vehicle drives them to Dr Gopal's office. The storm has finished, but it has left a trail of destruction. Power lines are down and there are several cars with smashed windscreens. Dark clouds hang menacingly in the air, threatening to unleash another deluge.

Dr Gopal meets them in his office. He wears a grave look and Melinda dreads what he is about to tell them.

'Thank you for coming down. I wanted to tell you in person,' Dr Gopal says. 'I'm afraid the foetuses have failed,'

'How? Why?' Cory asks angrily.

Melinda pushes past Gopal and walks out the back door of his office. Gopal explains as he hastily follows after her.

'We lost power during the storm.'

'And don't you have a backup generator?' Cory asks.

'Of course. But we used up all our fuel in the last storm. And it's practically impossible to buy fuel at the moment.'

'This is ridiculous!' Cory snaps.

Melinda remains silent as she flows down the corridor towards her unborn children. She enters the room and stares at each uterus. The monitoring panels have been turned off and it is eerily quiet.

There is no hum from the machines and no blood pumping to the organs.

'Turn them back on!' Melinda screams.

'There's no point,' Gopal says.

'Turn them back on!' she repeats.

'The foetuses are dead. There is no point having the machines on. I'm sorry.'

Melinda glares at Gopal with such venom that he takes a step backwards.

'Of course, right away,' he mumbles.

He fiddles with the machines and sounds fill the room, the pulsating rhythm of the blood machines and a continuous high-pitched tone from the display. All three heart rate monitors show a flat line.

Melinda collapses to her knees, great sobs wracking her body.

'This is on you, Gopal,' Cory says. 'I expect you to try again. But don't expect us to stump up three-million dollars again.'

Gopal nods. 'Operating costs only. I'll call and we can discuss the details.'

He ducks out of the room. Scurrying away before Melinda can direct any more of her rage at him.

Cory waits for Melinda to partially recover and then leads her back to the car. It is raining again and the foul weather makes both of their vile moods even worse.

They try to relax in the back of the automated vehicle, but Melinda can see the anger and pain eating away at Cory from the inside. Halfway home, the rain turns into a downpour and the windscreen wipers can barely keep up. The car flashes a warning and begins to slow down.

'We'd better pull over,' Melinda says. 'It's not safe in this weather.'

'It's only just started,' Cory says. 'If we pull over, we'll be stuck, and who knows how bad it'll get. I want to be home before the real storm kicks in.'

He taps away at the central control pad and overrides the car's safety precautions. They hurtle down the highway, passing rows of stationary cars.

'At least slow it down,' Melinda says, agitated. 'Cory, slow down!'

'We don't have time,' he says, his grief blurring his reason. He taps the control pad, encouraging the car to get them home even faster.

'You fucking idiot!' Melinda yells.

The rain falls as if it's a wall of water, specks of hail mingled with the liquid raindrops. The safety system can't handle the conditions with Cory ignoring its warnings and the vehicle slides elegantly towards the concrete barrier. They hit with such force that their bones shatter and their blood vessels pop. Their luxury car is reduced to a warped wreck and they are both dead within milliseconds.

SARAH

CHAPTER 20

As a community, we have a great breakthrough after six months when some of the former Zeta volunteers find a new water source close to camp. They are preparing a new area to quarry stone for more buildings and upon moving a large slab, find a pool of water.

We assume that it is fed by a natural underground spring, before seeping back into the ground and flowing as an underground river down the mountain.

The volunteers empty their canteens and fill them with fresh water, revelling in their find and the fact that they don't need to hike two kilometres to the river to do it.

I am sitting around the central firepit with a few other expectant mothers, when they rush into camp to tell us their news. Everyone else appreciates the find but can't match the enthusiasm of those who made the discovery.

I take a walk to stretch my legs and stroll to the farm area, which has become one of my favourite places in our village. I am joined by Meaghan and Poh, neither of whom have started to try for children yet.

I rest my hands on my bump as I walk, trying to remember how big my baby is supposed to be at five months. I'm pretty sure they should be as long as a banana.

'How does it feel?' Meaghan asks, noticing my hand position.

'Weirder than I expected,' I reply. 'As a midwife, I talked to women about being pregnant every working day but that didn't prepare me for actually experiencing it.'

'It's so exciting,' Poh says energetically. 'I think you'll make a great mother. And now Cassidy is pregnant again, your kids will be pretty close in age.'

'I know. I'm so happy for her,' I reply.

'Let's just hope she can hold onto this one,' Meaghan says.

I hold up my hand, showing my fingers firmly crossed.

As I place my hand back on my belly, baby starts to shift and wriggle inside my uterus. I flinch and pause, adjusting to the sensation.

'You alright?' Poh asks.

'Fine, baby's just letting me know they're here,' I say, smiling. I sit down on a tree stump, next to the protective farm wall. 'Oh, they're kicking.'

Poh's eyes light up in excitement. 'Can I feel?'

'You can try. Tan can't always feel it so don't be disappointed if you can't.'

Poh places her hand on my stomach and waits patiently. Just as I'm about to tell her that I think baby has stopped, they give my insides a sudden jab.

'Argh, I felt it. I felt it,' Poh says taking her hand away. 'Oh my god, that's so weird. But so amazing.'

'Try having it inside you,' I say with a laugh.

Meaghan has a go feeling for the kicks and baby obliges by giving a flurry of movements. Meaghan remains a lot calmer than Poh but smiles at me delightedly and places a gentle kiss on my forehead.

The following day, the medical team is stirred into action as a bug starts to pass through the town. At first, when Sylvie starts vomiting, we assume that she has fallen pregnant and is having her first morning sickness. However, her queasiness doesn't pass, and she continues to be sick at regular intervals. We try and keep her hydrated, but every time she has something to eat or drink, she vomits it right back up. Once Chris starts vomiting as well, we realise that we must've picked up a gastro bug from somewhere and the medicos jump to work checking on everybody. Once we start asking the right questions, we find that lots of people have been experiencing diarrhea but were too embarrassed to tell anyone.

As the day progresses and more people become violently ill, I try to work out the source of the infection. We move the sick volunteers to one of the large buildings so that we can look after them more easily. There are a dozen beds for the poorly to rest in, made from wooden boxes full of pine needles. Karim, Smithy, Ailish, Iris, Joyce, Latoya and a handful of former Zeta volunteers contract the illness and spend the rest of the day chucking up their guts. Luckily, Nia has crafted some buckets from melted chemo-pods, but we don't have enough to cope with the amount of sick the volunteers are producing. We have to improvise and use some of the larger, leftover food sachets as sick bags.

The sickroom smells awful and often one person vomiting triggers a round of vomiting from the other volunteers.

Edith tries to tell me that I shouldn't be nursing the poorly when I'm more than five months pregnant, but I remind her that without me and Cassidy, who is also pregnant, there would only be three medical staff available. She reluctantly concedes the point and lets me continue my job.

Sinclair, Tan, Edith, Cassidy and I struggle to keep everyone hydrated, but are also assisted by healthy volunteers who fetch water and cook hot soups for the poorly.

As Newton hands me a basket of water canteens, I have a sudden realisation.

'Where did you fill these from?' I ask.

'The spring. Why?' he replies.

'Tip them out. Fetch water from the river and boil these in it before you reuse them,' I say, gesturing to the canteens. 'And anything else that has had spring water in it.'

His eyes widen as he understands what I'm saying.

'Have you drunk any spring water?' I ask him.

He shakes his head. 'My canteen was still full of river water.'

'Good, now get me some clean water as fast as you can.'

He races off, yelling orders at the other healthy volunteers.

I duck back inside the sick house and find Karim lying on a bed of pine needles.

'Did you drink spring water?' I ask him.

'Yeah, Sylvie gave me some of hers after they found it yesterday,' he says.

My forehead becomes furrowed and he looks at me nervously.

'Why?' he asks.

I leave the question unanswered as I rush to tell the others of my suspicions.

We expect the symptoms to lessen overnight but instead, they seem to get worse. We sleep in shifts and spend our waking hours helping volunteers rush to the toilet or emptying buckets of spew. We cordon off the toilet to the other volunteers and minimize the amount of people dealing with the sick. It becomes clear that our small supply of rubber gloves is not going to be enough, so we stop changing them over. This means that we're protected, but we will readily spread germs amongst our patients.

What I initially thought to be a simple gastro bug is turning into something much more dangerous. Despite our best efforts, by morning, most of the volunteers are severely dehydrated. They are unable to keep down any food or liquids, and we have no intravenous fluids available to us.

The skin on their hands becomes wrinkly as the dehydration worsens and some volunteers begin to look gaunt, with sunken eyes. All of them are exhausted. To make matters worse, Sinclair becomes ill himself and is unable to care for the others.

We make the decision to give all the sick patients antibiotics and antivirals and redouble our efforts to get them to consume fluids. Using fresh river water, we make a warm, almost flavourless broth and instruct the volunteers to sip it extremely slowly, preferably once a minute while holding it in their mouths and letting it trickle slowly down their throat.

Following our advice, some of the volunteers begin to take on some water, but soon Joyce becomes our first fatality. She starts to complain of cramps, her blood pressure drops severely, and her breathing becomes laboured before she passes into a coma. We try and save her, but with limited technology, we ultimately watch helplessly as her skin turns blueish-grey and eventually her heart stops completely.

We are shocked and left without answers as to how to stop the other volunteers reaching the same fate. Already, many of them are exhibiting advanced complications.

When I tell Iris that Joyce has passed away, she manages to smile and make light of the situation.

'Hey, you managed to tell us apart correctly,' she says with a weak laugh.

'Well, you'd hope that I could after six months,' I reply.

'Only last week you called me Joyce,' she chuckles.

'Did I?'

She nods.

'I'm very sorry,' I say solemnly.

'Don't worry about it, Zuhra,' she says before poking her tongue out at me.

I hope that her ability to have a laugh is a positive sign, but half an hour later, Iris herself slips peacefully into a coma. A few minutes later, she passes away, taking her young foetus with her.

Two of the original Zeta volunteers have also died, one after having violent seizures, and we decide that we need to tell the healthy volunteers what's going on.

'Can we see them?' Yi-Seo asks, after we've delivered the bad news.

'That might not be the best idea,' Edith says. 'We still don't know for certain how the bug is spread.'

'But you guys haven't got sick,' Yi-Seo says. 'We should be okay to just talk to our friends and pay our respects to the dead.'

'At your own risk,' I say calmly. 'None of us know if we're going to get ill, but we've decided their care is important. You enter accepting the risks yourselves. And if you do, don't touch anything.'

Yi-Seo nods her understanding and brushes past me as she enters the sick house. Newton and a few of the other volunteers also come inside, but many of the healthy volunteers stay well clear, watching the building nervously. I can't begrudge them that.

I duck back inside, and the smell of death and sick smacks me in the face. I breathe through the discomfort and get back to work as I slowly acclimatise to the stench.

I feel worthless, flitting from makeshift bed to bed, topping up water cups, emptying sick bags but unable to offer any other meaningful care.

Newton is crouched next to Ailish. She is trying to talk to him but is struggling as her breathing becomes laboured and painful.

'You'll have to... carry the ginger baton... alone,' she manages.

He smiles at her. 'I wouldn't give up hope on you that quickly. You're the strong one.'

'Not this time,' she says, shaking her head. 'And tell Dart... I love him. I don't... see him in here.'

'He's waiting outside,' I say gently. 'I'll make sure he gets your message.'

I check her pulse, which is rapid and she starts mumbling nonsense as she passes in and out of consciousness.

Newton leaves, unable to watch her leaving our world. I stay with her until she has passed and then move to the next volunteer, hoping for some positive news.

I find Smithy, sleeping peacefully in his bed. I check his pulse, which seems steady and leave him to rest.

I spot Cassidy kneeling next to Karim, weeping silent tears. I squat next to her and look down at my strong, brilliant friend, lying motionless in front of me. The illness has shrivelled his skin and made him look a gaunt version of the vibrant man we knew. I thought he would look peaceful in death, but he just looks broken. Cassidy's tears fall onto his face and leave streaks where they cut through the grime.

I feel fear, sorrow and anger rising from my stomach and into my throat, but I push it back down. I need to remain strong for Cassidy, and there will be time later. Time to rage and time to cry.

I hold her to me as she starts to sob and I run my hand through her hair. We stay huddled together on the ground as she cries herself out. I feel guilty, taking time away from nursing the other sick, but I also know that there's nothing we can do. At this point, we must wait out the storm and deal with the carnage afterwards.

Edith and Tan continue to care for the ever-dwindling number of patients. I help Cassidy to stand up and lead her outside. I take a few deep breaths of fresh air before returning inside, leaving Cassidy with Newton.

Once most of the sick volunteers have passed away, Tan, Edith and I stand vigil beside Sinclair's bed. He looks up at us with fear in his eyes. As a nurse, he knows what to expect more than the others. I hold his hand and put a cup to his lips, but he pushes it away.

'If I have to go, I'd like to go… without vomiting any more times,' he says.

I nod gently and step back, but he doesn't let go of my hand. I hold it tightly, squeezing it supportively.

'Did you drink any of the spring water?' I ask while we wait.

He nods his head lethargically and has to pause every few words. 'Someone brought some in… when we were trying…trying to rehydrate everyone… yesterday. I was… thirsty… so had a few mouthfuls… myself.'

'I'm so sorry.'

He shrugs, resigned to his fate.

Sinclair holds on for another couple of hours before his arm finally becomes slack in my grasp. I place his shrivelled hand beside him and take a step back from his bed.

I'm shattered and all I want to do is collapse in a heap and cry, but I know I have to last just a few moments longer.

In total, twelve volunteers have left the mortal realm and their bodies lie on their pine needle beds. Many of them soiled themselves after they were too weak to use the toilet, and the heady odour of the room makes my eyes sting.

Smithy still sleeps in his bed, and Emmett also appears to have pulled through the worst of the illness. I'm sure that they both must've drunk the spring water, but for some reason, their bodies are better able to fight the illness.

We move them both to a smaller house and leave the dead volunteers to their final rest.

Edith offers to keep an eye on Smithy and Emmett.

I am about to head to my own house but feel an overwhelming need to feel clean. I let my feet take me down the mountain to the river. I walk in a fog of sorrow and exhaustion and arrive without noticing anything along the way. I strip off all my clothes and, sitting in the shallows, finally allow myself to cry.

I don't know how long I stay there, trying to wash away the dirt and the feeling of guilt, but the sky has started to darken by the time Tan finds me. He takes off his own shoes and trousers and wades into the water. I let him help me up, and he leads me to the bank. Only then do I realise that I'm shivering uncontrollably.

He wraps me in a blanket, and we sit down in the nook of a tree, his arms and legs holding me tight. He doesn't say anything, but he doesn't need to. He's here for me.

I put my clothes back on and we stumble up the slope, back to camp. Trees keep appearing out of the gloom, but by following the rope of The Machine, we find our way home.

I collapse into my bed of pine needles, wanting to weep for my friends, the brave souls who wanted a better world, but I'm completely out of tears.

Finally, sleep claims me. Mentally and physically exhausted, my slumber is deep and long, and not even a thunderstorm could wake me.

When I come to in the morning, I feel guilty for having left Edith and Tan to take care of the last ill patients on their own. It is Cassidy, however, that I find sitting next to Smithy's bed. When I enter the small house, she rises to meet me and engulfs me in an embrace. We cling to each other for a few moments before pulling apart.

'How are they?' I ask.

'Really good,' she replies, a hint of relief in her voice. 'They've both drunk plenty overnight and are keeping it down.'

'Brilliant. How long have you been on duty?'

She looks weary. 'Since early this morning,' she replies.

'Go get some rest, I can keep an eye on them.'

We embrace once more, not talking about Karim and the others but understanding each other's pain. Cassidy leaves and I take her spot next to Smithy. He has his eyes open and he's watching me closely.

'Where's everyone else?' he asks.

'The others who were sick?' I clarify.

He nods.

Zuhra enters the house and sits down beside me. She looks at Smithy tenderly.

'They're gone,' I say gently.

'All of them?' he asks.

'Everyone except you and Emmett,' I say sadly.

'It was the water, wasn't it?' he asks.

'We think so,' Zuhra replies. 'All of the other Zetas got sick, plus a handful of others that probably had some.'

'Well, I definitely tried it,' Smithy says glumly.

Zuhra shakes her head dejectedly. 'Rule number seven,' she mutters.

'I know, I know,' Smithy says. 'We got complacent.'

'We got stupid,' she replies.

'I can't argue that,' I say.

Smithy closes his eyes, the conversation taking a toll on his low energy stores. A few moments pass in silence before Zuhra speaks.

'I just can't understand why Allah would let this happen,' she says.

'As a test or a trial?' I suggest.

She ponders my response.

'I don't mean to be insensitive, but I can't quite understand how you can be religious at all in this new world,' I tell her.

'Why's that?' she asks.

'Well, firstly, why did he let the world die in the first place, and how is it okay for your parents to play God and create the R.E.Generation Project?'

'The answer is in your question,' she says calmly. 'If Allah enabled my parents to be able to create a way to chemo-preserve humans, then he wanted us to be able to do that. Allah used the Great Flood to purge the world, but humans did it themselves the second time around. The R.E.Generation Project is just the Ark to see humanity through.'

'I'd never thought about it like that,' I say humbly.

'Which brings me back to my first question,' Zuhra says. 'Why, after he delivered us here, would Allah allow half of us to die?'

'I only went to church for a few years as a child,' I tell her. 'But I would argue free will. It was our decision to ignore our training and drink the spring water.'

She shakes her head. 'The amount of times I've used that explanation to answer other people's questions… it doesn't really help hearing it from this side.'

'Yeah. I know what you mean,' I say.

I leave her to ponder her religious ideologies and watch over Smithy. I check on Emmett, who is still sleeping, but seems fine. His skin has returned to normal and his pulse is steady, indicating the worst of his dehydration has passed.

As I watch over Emmett, Meaghan and Poh tiptoe into the building and make their way to my side.

'Are you alright?' Meaghan asks in a hushed voice.

'As good as I can be, considering,' I reply. 'You guys?'

'Same,' Meaghan replies.

'It's just so unbelievable,' Poh adds. 'I have to keep reminding myself that it actually happened.'

'For a millisecond, when I woke up this morning, it was like any other morning,' I say. 'And then it all came flooding back and it was suffocating.'

Poh steps forward and pulls me into a hug. Meaghan asks her next question while we're still embracing.

'We were wondering what we do next?'

'Next?' I ask. 'I'm still worried about now.'

'I know, I know,' Meaghan says, as Poh lets me go. 'But we've lost too many volunteers. We need more.'

I glance at Emmett, who sleeps restfully beside us. They're right of course. We have the infrastructure now for thirty plus people, and we need fresh blood and DNA coming into our society.

'Are you suggesting what I think you are?' I ask.

'We need to get the others,' Meaghan says simply.

'They're not all bad like Declan,' Poh adds. 'Ayesha, Tomohiro, Jay. If anything, it was wrong of us to leave them under his thumb without anyone to oppose him.'

'There's also the large pod building full of chemo-pods. We could get more volunteers from there,' I say, trying to find a way to avoid Declan.

'Except that the computers seemed fried and we have no idea how to wake them up,' Meaghan says.

'And we really should try our friends first,' Poh adds.

'Friends?' I ask, whispering to avoid disturbing Emmett and hoping Zuhra can't overhear.

'You may have been in your own Eta bubble,' Poh replies. 'But you have to remember, I trained with Tiffany. She's a lovely person. Plus, Jay and I were developing quite a friendship. Who knows what would've happened if I'd stayed? It wasn't an easy choice for me to come with you, I hope you know that.'

'You never mentioned it,' I say humbly.

'I never questioned my decision once I was with you. You guys made me feel safe,' she says, her eyes becoming glassy. 'Now I want to provide that feeling to the others.'

I nod. 'You're right. We have unfinished business.' Meaghan smiles faintly in agreement and Poh touches my hand gently in thanks. 'But first we also have some unfinished business here.'

Miraculously, the entire council survived the horrific events of the last few days. I leave Meaghan and Poh to watch over the recovering pair and call a meeting with Zuhra, Tan and Graeme.

In regards to the dead volunteers, we decide that the best course of action is to burn the entire building, with everyone inside. It is a large wooden structure, one of our original houses, erected hastily to provide shelter for a large group of people. Once plans have been finalised for its incineration, I tell them that we need to go back to our original camp.

'There are volunteers out there that can help us,' I say. 'And it's also our responsibility to help them.'

'You've always said that you never want to see Declan again,' Tan says, confused.

'That's true. But I've been avoiding my problems, hiding here with you. I think it's time to face them head on.'

'It doesn't make sense,' Graeme says. 'We have everything we need here. We're safe from the storms.'

'We have everything we need here, except enough people,' I argue.

'There is no specific rule that specifies this, but I agree with the boys,' Zuhra says. 'You're taking a risk without enough of a reward.'

'They're right,' Tan says. 'Your journey here wasn't the easiest, and I don't want to risk the lives of my people so that you can settle an old score.'

'That's not what this is about,' I say defensively. 'It wasn't even my idea.'

'I want to talk to the other Deltas,' Tan says. 'I think it's important everyone has a say in this matter.'

'Okay, fair enough,' I say with a sigh. 'Meet back here?'

They all nod. Tan goes to round up the Delta volunteers and Graeme gathers what remains of the Beta and Theta volunteers. Zuhra takes a few steps away to go find the Zeta volunteers before she remembers she is the only one left.

Cassidy is still sleeping but I find Newton and we meet up with the girls beside Smithy's sick bed.

'I think we should go too,' Newton says. 'I've had a niggling feeling nestled away in the back of my brain ever since we left. I couldn't have told you what it was, but I think maybe it's a bit of lingering guilt.'

'If Zuhra doesn't go, I won't go,' Smithy says simply.

'She can come too,' I say hopefully. 'Her survival skills could be very useful.'

'I'll leave the decision to her,' Smithy replies. 'But whatever she does, I do.'

I nod my understanding. 'Thanks, guys, I'll see what the others have decided.'

Tan and Graeme take longer to debate the issue with their groups, but after some time, they join me and Zuhra at the central fire pit.

Graeme is the first to speak. 'All of the Beta and Theta volunteers are going to stay. If you insist on going, we recommend that you return as soon as you can. We will keep the farm running and provide a safe place for you to return to.'

'Thank you, Graeme,' I say, placing my hand on his warmly. 'I understand.'

'The Delta volunteers agree with Graeme,' Tan says. 'And as their leader, I will be staying with them.'

'Okay, but we have to go,' I say simply.

'As your partner I need to ask you, but I will ask you only once,' Tan says. 'Please stay. Let the others go, but you stay. After all, you're nearly six months pregnant.'

'We will only be gone a month,' I say. 'Two weeks there and two weeks back. I'm sorry, Tan, but I have to go.'

'I respect your decision,' he says calmly.

'And I respect yours,' I reply.

'This is bullshit,' Zuhra says. 'Pardon the language. Rule four hundred and three: for the greatest chance of survival, always keep all survivors together.'

'I don't know what else to say, Zuhra.' I hold my hands out apologetically. 'We're going and we can't force you to come with us. So, we'll have to split up for a short while.'

'Tan's right, you simply can't go while you're six months pregnant. It's not worth the risk.' She is almost pleading with me.

'It will be good to be active,' I reply. 'Plus, the others are my responsibility. I owe them all too much.'

'I don't agree with this,' Zuhra says stubbornly.

'I get that, but it's also not going to change anything, so let's all just move on, shall we?'

We spend the afternoon chopping down some trees and collecting dried pine needles, to fill the sick house. We lay lots of extra fuel inside, covering the dead bodies, and leaving anything that we used to treat them inside. Blankets, buckets and food wrappers will all be consumed by flames. We don't want to take any risks with whatever bacteria or virus killed our friends.

Many of the volunteers are still uncomfortable coming inside, so they lay firewood and kindling around the outside wall of the building.

Cassidy prepares twenty branches to use as torches and stacks them in a pile next to the central fireplace.

Emmett and Smithy are strong enough to join us for dinner and the remaining volunteers all share a meal in honour of our fallen comrades. To begin with, we are very sombre, but as we start to talk about the dead volunteers, the mood becomes inexplicably lighter. We share stories of them that make us laugh and remember the good times that we've shared together over the past six months. However, the levity is short lived and as the meal comes to an end, we become quiet and solemn once more.

The sun has set during dinner, leaving the fire as our only light source. One by one, we select a branch from Cassidy's pile and stand around the fire. Once everyone is ready, we place our torches in the flames, wait for them to catch and walk in a procession towards the sick house. Newton carries Callum's torch so that he can wheel himself to the house in his chair.

We stand in a large oval around the building and place our torches in the dry kindling gathered around the base of the wall. The flames lick up the side of the house and I have to take a few steps back as heat radiates from the blaze and the night is filled with brilliant light.

Dart stands to my left, tears streaming down his face. I stand next to him and hold his hand.

'Ailish told us to tell you that she loved you,' I say.

'I know. Newton told me,' he says. He tries unsuccessfully to blink away his tears. 'I just wish she'd told me when she was alive, so that I could've told her I loved her too.'

He is wracked with silent sobs, and Nia moves closer on his other side, taking his other hand in hers. Other volunteers move to us and we form a large chain, holding hands and watching the fire engulf the building. We stand as one until the roof caves in. We begin to break apart, led by Dart, who takes himself off to his house.

'I'll look after him,' Nia says, following after Dart.

I stand, continuing to watch the fire burn, unable to pull myself away. Cassidy drapes herself over my shoulders and I grasp her hand protectively. Tan takes a step towards me, changes his mind and walks away.

Cassidy and I, along with some of the other volunteers, stay with the fire late into the night. We only retire to bed once the last flicker of flame dies, leaving a huge pile of smouldering coals and ashes.

MAYUKA

As population declined, property prices plummeted and cities shrank. Those who survived were suddenly able to afford inner city houses and the outer suburbs of cities all over the world became ghost towns. Ghost towns with plenty of ghosts.

Mayuka darts from one building to the next, sticking to the shadows. This serves the double purpose of avoiding other ghosts and staying out of the harsh sun. Her white wet suit doesn't assist with stealth, but the bright material does reflect the sun's heat when she is forced out into the open and insulates her from the oppressive air. She breathes through a respirator, ensuring that she inhales cool air to maintain a safe body temperature.

She angles towards a shopping mall. It has been her go to location for food for the last month, but she fears its supplies are almost exhausted. She passes through the front door, remembering the days when that meant walking into air-conditioned bliss. Now, the interior of the mall is as humid and hot as outside. She avoids the supermarkets – they were the first places to be plundered and she knows there is nothing worthwhile left. Hopefully the pet shop still has supplies. The last time that she visited, she left the security gate closed and so her heart sinks when she sees it open. The entire store has been ransacked and the only thing she can find is one dented tin of dog food. She finds it funny to think that people once wasted perfectly good food on pets. She places the tin in her bag and continues her search.

She manages to find dregs in a few other stores, long-life milk at the donut shop, some lollies at a discount store, but the mall has been bled almost dry.

She heads for the exit and is confronted by three other ghosts at the door. They all wear white, but she notes that none of them have the luxury of a wet suit.

'Well, well, well,' the middle ghost says. 'Looks like a little mouse has been sneaking around and stealing *our* food. Hand it over, little mouse.'

Mayuka doesn't have the time or patience for this. She surprises the ghosts by sprinting straight towards them and they act on instinct, flinching away from her mad dash. She pounces on the opening and darts between them before they can react. She races outside and they give chase, wasting energy by yelling curses at her.

She sprints down the middle of the street, in broad daylight and fully exposed to the sun. Such physical exertion in the heat of the day is almost a death sentence, but she knows that her pursuers will suffer even more than her. After about a kilometre, she darts inside a house to wait. She breathes heavily as her body heat is trapped inside the wet suit and threatens to overwhelm her. She heads for the shower and is relieved to find the water connected. The water is warm at first, but still cooler than her and the air, so she lets it wash over her. She strips the wet suit to her waist and stands motionless, enjoying the feel of the water against her body. She lets the water flow into her mouth and swallows a few mouthfuls. The other ghosts have clearly given up the chase, so she takes her time to enjoy the shower and get her body temperature back to a safe level.

Once she is finished, she puts her arms into the sleeves of the wetsuit and zips it closed. She doesn't bother drying off, because she'll be dry in a few minutes anyway. She checks the house for food but, finding nothing, she leaves and carefully makes her way towards her own dwelling. When she had run from the mall, she made sure she didn't run towards her house, so she now has to swing around to get home.

She breathes a sigh of relief as she climbs through the window and stands in her kitchen. She deliberately chose a house that looked run down, to limit the risk of other ghosts invading. The front wall of the house is littered with bullet holes and police tape criss-crosses the front door. But what the rough exterior hides is a wonderful inner sanctum. An airtight lounge room fitted with an air conditioner connected directly to solar panels on the roof. Mayuka strips off her wet suit and respirator and collapses onto the couch, enjoying the cool air on her skin.

She still feels a bit under the weather after her mad dash, so she spends the afternoon lying on the couch, mulling over her situation. She will need to go foraging for more food tomorrow, but she can't think of a good place to get supplies.

She pulls the tin of dog food out of her bag and opens it for dinner. It has a nice gravy and large chunks of meat. She can't be certain, but it even appears to have some peas in it which would be the first vegetables she's eaten in several days.

An hour later, she doesn't enjoy the flavours nearly as much as she regurgitates the food into the toilet. She wipes her mouth and stands up, which brings on another bought of retching. Unfortunately, there is nothing in her stomach to throw up except a small amount of bile.

She crawls to the kitchen and manages to keep down a glass of water. After grabbing a bucket, she heads back to the couch and sleeps well until a bit after midnight when she wakes and rushes to the toilet. This time she has severe diarrhoea, and when she tries to drink more water, she brings it straight back up. After a few attempts, she manages to keep down some out-of-date paracetamol and antiemetics, taking them with tiny sips of water. Mayuka had squirreled the drugs away years ago and had hoped to never need them.

She retreats to the couch once more, feeling horrible. She doesn't think that the dog food could have made her feel this way, so she searches for what else could've inflicted her. She has always drunk the water at her house without any problems, but she deduces the water she showered in must come from a different source. She curses her bad luck as she drifts in and out of sleep and her condition slowly worsens.

Mayuka battles the illness for the next thirty six hours, but she is unable to consume enough fluids to sustain her body and she passes into a coma. The air conditioner continues to pump cool air into the lounge, but without anyone to care for her, it isn't long before Mayuka's ghost leaves her broken body.

SARAH

CHAPTER 21

The following morning, I have a long sleep in. Although we are keen to begin our journey back to our original base, we have decided to wait until after the next storm. After we woke the volunteers from Unit Two, the storms became more frequent and at one point, we had to hunker down almost weekly. Over recent months, however, the gap between storms has widened once more, corresponding with the days getting shorter. Because of this, I assume we're in the middle of winter, or more pertinently, the dry season. While daily temperatures have dipped from their peak, it still doesn't get particularly cold. Considering the tropical weather here, in an area that used to be temperate, I hate to think what the weather is like in areas closer to the equator. I imagine the old tropical areas would be unbearably hot.

Once I get up for the day, I help with preparations for our hike. As we are waiting for the next storm, we have some spare time to work with, so we decide to design a wagon for Aravis and Tumnus to pull. It is a simple cart design, with wooden wheels and high walls to stop our supplies from falling out on rough terrain.

Zuhra's decision to stay at camp means that Smithy will not be joining us on the journey. As we're expecting the journey to be a round trip, we only need limited supplies. Water, food, half-empty oxygen tanks and tools in case we need to build another bridge, are all loaded into the wagon.

A few days later, Newton tells us that a storm is on its way. He has found a new weather tree at this camp and has continued to climb it at least once a day.

Tan and I shelter from the storm inside our house. Our relationship has been strained ever since I told him that I was leaving, with neither of us saying much to the other. As the rain lashes against the roof and the wind whistles around our small retreat, we find comfort in each other's presence, even without any words being spoken.

As I lie in bed, he places a hand and his head on my tummy. Baby seems to be sleeping, but Tan talks to them soothingly, wanting baby to recognise his voice.

Tan lies down behind me, leaving a hand on my belly.

'Do you really have to leave?' he whispers.

I nod slowly.

'I don't want you to. I want you to stay with me,' he says lovingly.

'I know. I want you to come with me,' I say.

'I know,' he says with a big sigh. 'But I'm needed here.'

'That you are. Keep everyone safe and make sure you're here when I get back, okay?'

'Of course.' He kisses the back of my head. 'Make sure you come back.'

'Of course. But if the unthinkable happens and we don't come back–'

'I don't want to think about that.'

'If we don't come back, there's something you should know.' I roll over to face him. 'We found a building full of newer chemo-pods.'

If, for some reason, we can't get back to them, it's important that they know where they can find more chemo-preserved volunteers. He is shocked that I've been keeping a secret from him, but grateful for the information.

We discuss more plans for the time that I'm away and eventually drift off to sleep.

The next day, the storm has blown itself out and we make our final checks for our journey. Bobby, our resident tanner, has made a harness out of snake leather to fit the lizards. Now about the same size as crocodiles, Aravis and Tumnus are muscular and powerful. Aravis allows me to slip the harness over her head and tighten it around her middle. I attach the wagon and she looks at me excitedly. We have practiced with an empty wagon a few times over the last few days and she seems to really enjoy pulling it behind her.

We hug the other volunteers goodbye, still not used to seeing the group with such a large contingent missing. Five of us follow Aravis and Tumnus down towards the river, leaving only fourteen volunteers waving goodbye.

For the next three days, we follow Meaghan as she leads us towards the lake. On her expeditions to collect snake meat, she has determined a more direct route that doesn't follow the curves of the river, shortening the distance so that the walk takes three days instead of four. Initially, I think the hike is good for me, but after a few days, I notice pain in my pelvic girdle. I grit my teeth, hoping it doesn't worsen as my ligaments continue to relax through the pregnancy.

We reach the lake on the afternoon of our third day and enjoy a dip in the shallows before the sun sets. Most of our supplies are in the wagon being pulled by Aravis and Tumnus, which has made the hike easier and more enjoyable.

Meaghan kills two large snakes and we enjoy skewered chunks of meat cooked over a fire for dinner.

After organising the evening watch, we all turn in for the night. I make sure I find a spot close to the edge of the tree line so that I get a wonderful view of the starry night sky.

We continue around the lake, making good time. The long grasses of the lakeside plain and the blue water are pleasant company, but the long, silent hours provide too much time for me to contemplate and remember. I keep expecting to turn around and see Karim's laughing face, but I'll never see his dimpled cheeks again.

I think about the care that we provided to the sick, trying to work out if there was anything we could've done differently, but always come back to the same conclusion. With more resources and old technology, we might've saved them all, but with what we had available, we did the best we could. Ultimately, we didn't stand a chance.

Maybe we should've been more cautious with the water, but after having had no difficulties for six months, it makes sense that we became lax. Maybe we should've given antibiotics and antivirals sooner, but Sinclair was treated as soon as he became symptomatic and the drugs still didn't help him.

Whatever the pathogen was, bacteria, virus or something else entirely, it has had thousands of years to mutate and our medicine hasn't kept up.

Unfortunately, no matter how I justify it to myself it doesn't make the guilt or sorrow go away. Some days I feel like I'm still walking in fog, unable to take in the beautiful sights around me. My mission now is simple. One foot in front of the other. We couldn't save Karim, but maybe we can save Tomohiro and Ayesha. I have to hope that we can, otherwise I might be walking in fog forever.

We are all subdued on the trek, but Cassidy is the quietest of all. I worry about her mental health after the loss of her partner and resolve myself to check in with her. Karim was her rock, so she'll need someone else to lean on now.

We complete the journey around the lake without incident. Meaghan keeps us well fed with snake meat, and we enjoy some delicious meals, making stews and stir-fries with the vegetables that we brought with us.

Aravis and Tumnus seem naturally suited to pulling the wagon. They make no complaints and trot gayly alongside us, hauling our supplies. In fact, Aravis is most grumpy when I take the harness off her and strap it to Tumnus.

I really appreciate them taking the load, because my back aches more and more as the days progress. I try and maintain a good posture as we hike, but with a small, bulging belly I keep finding myself leaning backwards to compensate. My pain worsens day by day. It feels like my bones are grinding against each other and without a chance to rest, the pain slowly builds. Cassidy is two months behind me, but her belly is now starting to show. I check in with her but she says she feels fine. Physically, at least.

We reach a large river flowing out of the lake on its northern edge. We follow it away from the lake and come across the remnants of our bridge the next day. It is largely as we left it, with the two long poles straddling the bank on either side of the river.

Some debris has caught on the poles, presumably from when the water level was higher. The most spectacular is a small, dead tree that has wrapped itself around the bridge about halfway across.

We unload tools and planks of wood from the wagon. Meaghan checks that the poles are secured tightly, ties a rope around her waist and ventures out onto the bridge, hammering in planks as she goes to make the bridge functional once more. When she reaches the tree, I cautiously follow her onto the bridge and offer her a saw. The bridge bows menacingly under our combined weight, so I hastily retreat and leave her to deal with the tree independently. She is able to remove some of the branches that are snagged and shove the dead tree off so that it floats under the bridge. Meaghan completes the renovation and stands triumphantly on the opposite bank, hands on hips in a power pose.

We applaud her politely before making our way across the bridge. Tumnus and Cassidy cross together. Tumnus is surefooted but stays close to her for comfort. The bridge sags as they reach the middle but holds, the planks providing extra structural support. Once they've reached the opposite bank, I encourage Aravis, who is harnessed to the wagon, to follow them across. I walk behind her, guiding the wagon and ensuring that it doesn't fall off into the river. It bumps over the gaps between each plank, but the wheels are large enough to keep turning without getting stuck.

My heart jumps into my throat as the bridge suddenly jerks downwards, but the movement only lasts a second. The bridge settles into its new position and we continue to safety over to the other side of the river.

Poh follows tentatively, crawling across the bridge on all fours, and Newton brings up the rear. Thanks to Nia, we have plenty of nails now, so we leave the bridge constructed, waiting for our return journey.

'Well, that was easier than last time,' Meaghan says, smiling.

'Absolutely,' Poh adds.

I pat Aravis on the head and her forked tongue gently licks my wrist.

'Shall we continue on?' Newton asks. He pulls out the compass and directs us north-west. Up until this point, we have been able to navigate using the landmarks, but now some precise direction and good memory is needed to get us back to camp.

'You know what I've been thinking about?' Poh ponders as we press on.

'Do tell,' Meaghan probes.

'It's quite ironic that Newton is the last male,' Poh says.

'Why's that?' Newton asks.

'Hiking through the forest with four women. That sounds like the beginning of a fantasy for most guys,' she says. 'Not for Newton though of course.'

'Newton and his girls,' I add, smiling.

'It may be ironic, but it's probably a good thing,' Newton says. 'Avoiding all that drama, we can all just carry on as the best of friends.'

I smile at Poh, appreciating her ability to find a lighter side, while my thoughts are all still consumed by darkness.

We follow Newton and the compass for a few hours, fording smaller streams, before stopping for dinner. After we've eaten, Newton takes Meaghan to climb a tree to check on the landscape and ensure we're on the right track. I'm a little jealous. That used to be our thing.

When they get back, they report that we have almost reached the foothills of our own mountains and are heading for a wide valley that sweeps up towards the mountains proper. The valley should eventually lead us back towards our camp.

After a night's rest, we continue on our way, drawing ever closer to Declan and the other volunteers. We find a river that leads towards our valley and follow it upstream. The river slowly decreases in size as we cross tributaries that feed into it from the hillsides.

As we hike, Cassidy pulls me aside and we talk about what's going to happen when we meet up with the others.

'I need you to promise that you won't do anything stupid when you see Declan,' she says.

'I'm not sure I can make that promise,' I tell her truthfully. 'I honestly don't know how I'm going to react when I see that little shithead.'

'You're gonna react with decorum and civility,' Cassidy says.

I smile weakly. 'For everyone's sake, I hope you're right.'

'You're better than him,' she says stubbornly. 'Please remember you're better than him and keep calm. Don't let him under your skin. That's how you win.'

As the sun begins to set, we set up camp next to the water, which is now only a small stream. I massage my pubic bone, which is sore and achey. I'm looking forward to this trek being over. We have made good time the entire hike and should easily make it back to altitude before the next storm. Assuming that we can find the way back to our original camp, we are confident we'll arrive sometime the following day.

Early in our journey the next morning, Cassidy points out a dried creek bed stretching away from a bend in the stream. It has been filled with a lot of pine needles since we were last here, but the depression the flowing water made over hundreds of years is still visible, as are the occasional pools of water scattered along the old water course.

We follow the dead stream up hill, gaining height rapidly and panting as the hill climb starts to catch up with us. We all have our heads down, concentrating on the slope and so I walk straight into the back of the wagon when Aravis suddenly stops at the front of the group. Looking up, I see a large lizard blocking our path. It is bigger than both Aravis and Tumnus and is looking at our group quizzically, with its head tilted to one side. The other lizard waves its arm in a large sweeping motion and Aravis and Tumnus duck their heads, as if in respect.

Us humans just stand in awe, watching the beautiful beasts communicate in their strange way. They exchange a few more gestures before the other lizard abruptly turns and walks away.

Aravis turns to look at me.

'What was all that about?' I ask her.

She tilts her head to me, replicating her movement towards the large lizard, before turning away from me and continuing up the hill.

'Alright then,' I mutter under my breath.

'That was cool,' Poh says quietly. 'Weird, but cool.'

We continue uphill, the dry creek bed becoming harder to see, however, Aravis and Tumnus lead the way confidently. We crest a rise and I sigh in relief as I see a manmade trench. We have made it to the mountain spring.

The beginnings of the aqueduct carry the water away from us, down the hill and towards camp.

We take a short rest and refill our water bottles, but I don't want to linger too long. When I look at the spring, I can't help but see the spring at the new camp that killed eleven people and turned our world upside down. My mind plays tricks on me as I try and squash the nagging voices in my head. I even see small bacteria and viruses floating in the water, created and magnified by my distressed brain.

I tear myself away from the pool and start the walk downhill, towards camp. The others follow me without saying a word as we track beside the aqueduct.

The final seven kilometres seem like the longest distance of all. We have no idea what we're going to find. We don't know whether, in fact, these volunteers have survived and prospered in the months that we have been gone. More importantly, we don't know how they'll react to us returning. To deal with the uncertainty, we have devised a plan and decide to approach with caution. We could be walking into danger, and so we need to do some reconnaissance first.

Once we're close to camp, Newton and I leave the others and creep carefully forward. Aravis is unhappy to be left behind. She seems twitchy to get to camp for some reason, but I am able to calm her down and convince her to stay with Cassidy and Tumnus.

Newton and I use the built-up areas of the aqueduct to sneak closer to camp. We move excruciatingly slowly but eventually we get a glimpse into the clearing. They don't appear to have made many changes but thankfully we see volunteers alive and working. I'm surprised to see two unfamiliar faces, and even more shocked to see Mabel, one of the original Eta volunteers that I trained alongside. I mentally said goodbye to her months ago, even though we never found her pod.

I can't see Declan or Nat, but I glimpse Ayesha and Tiffany as they walk around the camp completing various jobs. The volunteers seem to be having lots of serious discussions while completing their daily tasks.

Ayesha walks towards us and we shrink back. I'm worried that she has spotted us, until I see a bucket in her hand and realise that she's coming to collect water. She is even more heavily pregnant than I am. She walks with one hand supporting her back as she struggles to move freely with her pregnant belly. She angles towards us and I make alarmed faces at Newton. He tries to show me a plan through some rapid gestures, but I shake my head, not understanding. He stands up slowly and disappears behind a tree.

I stay deadly still, hoping Ayesha doesn't spot me crouched next to the aqueduct. She is within ten metres of me when she stops to scoop water into her bucket. Newton materialises out of the forest, appearing behind Ayesha. He wraps one hand over her mouth and the other around her chest, pinning her arms to her sides. She starts to struggle, and I step forward to help Newton contain her. The moment that she sees me, she stops fighting against Newton, her eyes wide in surprise. We pull her deeper into the forest, leaving her bucket spilling water onto the ground. She allows us to tie a strip of cloth around her mouth and lets us lead her meekly, Newton holding her arms behind her back. She doesn't resist and I think she tries to smile at me when we make eye contact. It's hard to tell with the gag in place.

'This wasn't the plan,' I hiss at Newton, but he just shrugs.

We reach the others and Ayesha scans their faces, a solitary tear falling down her cheek.

'Are you gonna call out?' Newton asks.

Ayesha shakes her head.

He unties the cloth from her mouth and lets her relax. She smiles and there is an awkward moment when no one quite knows what to say.

'Hiya,' Ayesha says, breaking the ice. 'What's been going on?'

AMAHLE

'Do you remember going outside?' Tallulah asks as they tuck into their entrée.

Amahle's chef has managed to procure some exclusive items, including Brie from France and grapes from Italy. Amahle and her friends sit in her dining hall. On one side of the table, a full glass wall looks out over the city. There is no movement out there, no living creatures. The humans that couldn't afford to live inside have all died out and the wealthy survive in air-conditioned luxury. Keeping on just enough workers to grow food and see to their daily needs.

'I do actually remember going outside,' Amahle says. 'My father told me that before I was six, we went outside regularly. I remember a few instances. I didn't like it though. The air clung to your skin and the heat was indescribable.'

'That sounds horrible,' Lateef says. 'I'm glad I didn't have to experience that.'

'I wish I had,' Tallulah muses. 'Just once. Don't get me wrong, I'm very happy in here, but I do wonder what it is like out there.'

'I just wish I was richer,' Junior says. 'One of my friends told me that he knows someone who flies all around the world, visiting all of the natural wonders and tourist attractions, and he's like the only one there.'

'How does he manage that?' Jenna asks. 'Does he have a space suit?'

Junior shakes his head. 'He has his own private jet and inside the jet, he has a land vehicle which can be deployed onto the ground. He never has to go outside and he gets to see all those wonders from his air-conditioned buggy.'

'Wow, that does sound pretty incredible,' Barney says.

Amahle's staff clear their plates and the others all thank her for sharing such exotic fare with them. It isn't long before the conversation inevitably returns to children.

'I know we're all too old,' Barney begins. 'But has anyone heard of anyone else successfully having children?'

'My yoga instructor's brother just had a child,' Jenna replies.

'And is it healthy?' Amahle asks.

'Yes and no. He was born without any genitals, but they think they should be able to create some for him.'

'Where did we go wrong?' Junior asks.

'Well, Tallulah and I decided to get married for a start,' Jenna says. 'So, we were never gonna have children without medical support.'

'Yes, but you did try,' Junior says. 'Just like Abebi and I tried for years and years. We got the doctors involved, went through cycle after cycle…' he trails off, wistfully thinking about what might have been.

'My doctor always told me it was my parents' fault, or their parents' fault,' Barney says.

'We do love to blame the older generation,' Amahle says. 'Maybe it's a blessing that we can't have children so that they don't blame us for the world we leave them with.'

The others ponder this while the main course is served.

'This is a new type of meat that has been created in a lab,' Amahle's chef announces. 'It is supposed to taste like a combination of beef and lamb. It is served with a pumpkin puree and capsicatoe sauce.'

'Thanks, Chef,' Amahle says.

She stands behind Amahle as they sample the meat, keen to hear their thoughts on the new recipe.

'It has a strange flavour,' Lateef says. 'It doesn't really taste like meat, does it?'

'I guess that's because it was made in a lab?' Barney muses.

'But I've had lab-grown beef lots of times and it's delicious,' Lateef says. 'This has a really odd flavour. I can't quite put my finger on it.'

'I know what you mean,' Jenna says. 'It's quite bitter.'

'I'm sorry, Chef, but I don't think we like it very much,' Amahle says.

'Not to worry, it will be the last time I cook it for you,' she replies simply.

They all try the sauce and puree and find that, by mixing the flavours, they can mask some of the meat's bitterness.

'Tallulah, is that a rash on your neck?' Barney asks.

Tallulah pats at her throat. 'You know me, allergic to everything.' She pulls out an adrenaline syringe and stabs it into her thigh. 'I'll be fine,' she says.

'My throat feels a bit itchy too,' Lateef says. 'But on the inside.'

'I know what you mean,' Amahle says. 'I thought it was just something to do with the texture of the meat, but…' she trails off, her voice becoming raspy as her larynx constricts.

She starts to have difficulty focussing as air flow is restricted to her lungs, but she can see her guests struggling as well. She collapses off her chair and stares blankly at the ceiling. Her chef steps into view, looking down at her passively.

'Air…ambulance,' Amahle groans.

'I don't think so,' Chef says bitterly. 'The rich shall fall,' he adds as he steps out of Amahle's line of sight.

The six of them writhe on the floor, struggling to breathe, until one by one, they lose the fight with the poison and become totally silent.

SARAH

CHAPTER 22

We all look at Ayesha, not really sure how to answer her question.

Poh is the first to speak. 'Where do we start? There's too much to say, really.'

'What about here?' I ask. 'Are you guys alright?'

'Yes and no. But I sure am glad to see you guys,' Ayesha replies.

'What do you mean?' Cassidy asks.

'We've got a situation developing,' she says. Her demeanour suddenly becoming serious and sombre. 'Tomohiro's dead.'

'Dead. Tomohiro?' Newton asks, disbelievingly. 'He's as tough as nails. I thought he was gonna live forever.'

Ayesha nods sadly.

'What happened?' Meaghan asks.

'We found him with a knife in his back,' Ayesha explains. 'We've got Declan and Nat contained inside their hut but there's no evidence pinning it to them.'

'How did you manage that?' I ask. 'I didn't think it was possible to contain Declan anywhere.'

She raises her eyebrows at me. 'Group effort. Also, he's denying any involvement and knows we can't keep him indefinitely. Boy, is he going to be surprised to see you.'

'When did this happen?' Meaghan asks.

'The day before yesterday,' Ayesha replies. 'Some of the other volunteers are getting a bit twitchy. No one knows who to trust.'

'After the first murder in a new society, I can't say I'm surprised,' Poh says. 'Shit.'

'We'll have to have a trial,' Meaghan says.

'But there's no evidence of who did it?' Newton says. 'That's what Ayesha just said.'

'But there's lots of circumstantial evidence,' I say. 'It's ages ago now, but Declan and Nat literally asked me if I would be okay with euthanising Tomohiro. Of course I said no, but if that's not an admission of guilt, I don't know what is.'

'It was definitely Declan,' Ayesha says, regaining our attention. 'The man is a snake and totally evil. Plus, Tomohiro told me multiple times that he felt unsafe. That he thought Declan was going to off him.'

'It's not for us to say,' Meaghan says. 'Which is why we have a trial and get everything out in the open.'

Everyone nods agreement.

'I guess congratulations are in order, too,' Cassidy says, gesturing towards Ayesha's pregnant belly.

'Not really,' Ayesha replies. 'It's his. But still, a baby is a precious thing. I see you're expecting too, Sarah. Is it yours?' she asks, turning to Newton.

'No,' he says with a broad grin.

'Karim, then,' Ayesha says. 'Where is he anyway?'

Cassidy turns away at the mention of his name.

'Not Karim either,' I tell her.

'Smithy?' she asks.

I shake my head.

Her forehead furrows in confusion, before her eyes widen as she guesses at the only other possibility.

We briefly explain what we've been up to over the past seven months. She is silent as we explain the outbreak that claimed a third of our volunteers, including Karim, and nods as we suggest that we have come to gather the remaining volunteers before returning.

'We'll come with you,' she says simply. 'It's been bloody lonely here, to tell you the truth. We just have to deal with Tomohiro's death first. Declan will probably argue against going, but I don't think I'm the only one who is sick of him.'

I nod, encouraged by Ayesha's words.

'I should've come with you before,' Ayesha continues. 'Tomohiro told me you guys were leaving, but I didn't want to believe him and I was too scared of Declan.'

'And you're not scared of him anymore?' I ask.

'I'm terrified,' she responds. 'But you guys are here now. I've got hope again.'

I pull her into a hug and the others join in, holding her and each other in a large group embrace.

We disengage and I turn to Meaghan. 'So, how does this work then?'

We follow Ayesha back to camp, and she introduces us to the new members of her tribe. After we left, they were desperate for more volunteers, so they searched harder for the missing pods. Two volunteers were unearthed in the shallow ground of the landslide. Instead of being crushed in the deeper earth, as we had expected, they had been pushed a few hundred metres away and lightly buried. Mabel's pod was found high in a tree, invisible from the ground. It was surprisingly undamaged, but a complicated retrieval operation was required to get it to the ground safely.

Mabel has a soft, rounded face with big eyes. She is one of the sweetest women that you'd ever meet and looking at her, you would have no idea that she has a black belt in taekwondo. She's a trained cook and has years of experience weaving any and every material.

Todd and Kirra are both Gamma volunteers and Poh embraces them warmly when we arrive. Based purely on appearances, they couldn't be any more different. Kirra is a historian, beekeeper and chandler. She is slender, has dark skin and has maintained her appearance after six months of survival. She has made some make-up from coloured rocks that she crushed up and her hair is neat and tidy. Todd, too, has dark skin, but has grown out his long brown locks and has scruffy facial hair sprouting every which way. Like Tan, he isn't beautiful. He is built like a bulldog, and to be honest, looks a bit like one too. He has a ridiculous English accent that sounds put on, but the others assure us it's legitimate. He is a botanist and horticulturist and has been working with Ayesha to grow food for the group.

'Why didn't you tell us that you had seeds?' I ask Ayesha, confronting her about her deception.

'Why do you think?' she replies. 'Declan. He made me plant them in a secret location. Wanted to keep all the food for the Alphas. It was when I tried to argue against him that he raped me for the first time.'

The casualness with which she says that makes me realise how normal that is in her life now. Luckily for her, and all of us, that's about to change.

'I'm sorry to hear that,' Meaghan says.

I see Poh looking at her hands, probably recounting her own traumatic experience with Declan.

'I was always going to share the food with everyone, I just figured there was no point fighting him on the location,' Ayesha says. 'You guys left before anything had grown anyway.'

'I understand,' I tell her, placing a hand supportively on her shoulder. After seven months away, I'd almost forgotten just how horrible and controlling Declan can be.

We greet Jay and Tiffany politely, but we don't say much to each other. I get the impression that they are wary of us, unsure how our sudden reappearance is going to affect them. Tiffany's hair has grown significantly and is now two toned, with peroxide blonde ends and a significant amount of brown regrowth.

Meaghan takes charge, explaining that the following morning, Declan will face one charge of murder. She asks that if anyone else has any charges that they would like to lay against him, they inform her of these before the end of the night. Anyone who would like to give evidence will be able to, and due to the small nature of the group, everyone who was at camp at the time of the murder will be cross examined.

After all evidence has been presented, a decision will need to be reached by the jury, which in this case will be everyone, with a unanimous decision preferred.

We spend the rest of the evening getting reacquainted and sharing stories of our last seven months apart. I notice a few volunteers seeking Meaghan out and having private discussions. I don't visit Declan and he is kept contained inside his hut. The general mood is very nervy. The returned volunteers, myself included, are anxious about our reunion with Declan, while the others are trying to get our measure. Only Ayesha appears calm. She took on the leadership role after Tomohiro was found dead and Declan detained, but it weighed heavy on her. She appears relieved to have a forward path and to be able to hand the decision making over to us. She's more than happy to return to her role as a quiet observer.

When I wake up in the morning, I'm surprised to find Meaghan rearranging the log benches.

'You're up early, sleepy head,' I say cheekily. 'Normally you're still snoring at ten am if we don't wake you up.'

'Couldn't sleep,' she says, glancing up at me momentarily. 'Can you help with this please?'

She has arranged the seats into two lines, facing two wooden log stools. I help her shift the last bench into position and she sighs in satisfaction.

'You nervous?' I ask her.

'A little. After talking to some of the others, I've got a new understanding of who Declan is. This could go anywhere.'

I scrunch up my face.

'Let's get everyone ready,' Meaghan says, clapping me on the back.

An hour later, everyone except Declan, Nat, Todd and Jay have taken their seats. Meaghan sits uncomfortably on one of the exposed stools at the front of the group. Todd and Jay appear, marching Declan between them. His eyebrows shoot up when he sees me and the others, but he chooses to remain silent. They lead Declan to the other stool and he perches on it, back straight and highly alert. I can almost see the gears turning over inside his brain. Nat follows behind and finds a seat on an empty bench. She is heavily pregnant and looks extremely tired. Jay and Todd join her and an expectant silence hangs over the group.

'Nice of you to remember my birthday,' Declan says, making light of the situation and trying to disarm the gathered crowd.

Meaghan ignores him and begins. 'Declan McCallister – I'm told that's your full name – you are here to face several serious charges and your guilt will be decided by those present,' she says formally. 'The charges against you are as follows: one charge of murder, that of Tomohiro Hosoda three days ago; thirty seven charges of rape against two victims, over the course of the last eight months; and three cases of indecent assault against two further victims. Do you understand?'

Declan nods.

'Do you wish to enter a plea at this stage?' she asks.

'Not guilty on all accounts,' Declan replies. 'Obviously.'

'Alright. In that case, let's begin proceedings.'

Meaghan begins by asking each of the volunteers about the night of Tomohiro's murder to try and establish what happened. Ayesha and Tiffany recall hearing Tomohiro scream in pain and rushing to his hut. He had been living on his own in one of our original huts, and as such, no one else was around. They found Tomohiro slumped on the ground with a knife in his back and masses of blood loss. He was unconscious when they reached him and dead within minutes. They both report that Mabel was sleeping at the time and didn't wake up.

Jay and Todd did not hear Tomohiro scream and were only alerted when Ayesha came and found them. They claim that they were playing checkers together. Nat and Declan claim that they were inside their hut the entire night and, like the boys, were only alerted to the crime when Ayesha knocked on the door.

No one can provide any evidence that Declan wasn't in his hut, or that he was the one who killed Tomohiro. Many of them report having had discussions with Tomohiro where he had talked about his fear of Declan and that he felt threatened. This had particularly increased the last couple of weeks.

With no further evidence forthcoming about the murder charge, Meaghan turns her attention to the alleged rapes.

Poh begins by recalling her attack. She is shaking as she stands before the group, and I can tell she is incredibly nervous, but she shows tremendous courage outlining what happened. She doesn't look at Declan, but he watches her closely.

Next, Ayesha stands and accuses Declan of raping her thirty six times. She has recorded each instance in the back of her botanical notebook.

'The first ten entries are retrospective,' she admits. 'I only thought to start recording the events after Sarah accused Declan of trying to rape her. After Declan found out I was pregnant, the rate of attacks decreased, but never stopped completely. He wanted me to know that he could touch me whenever he wanted.' She shudders and I sense Cassidy tensing up beside me.

I glance across at Nat, and she watches proceedings with a stony composure. It's impossible to know how much of this she already knew.

Ayesha finishes her testimony and sits down beside me. I squeeze her shoulder supportively and she smiles at me weakly.

Next it's my turn and I give a brief account of Declan's attack on me inside the Shit House. Most of them know the details already, and it seems almost inconsequential in comparison to Ayesha's sustained abuse and Poh's attack but Meaghan insists I continue to 'get a complete picture'. I also recount my conversation with Declan and Nat, where they suggested killing Tomohiro because they thought he was a burden on our society.

Next up, Mabel also gives evidence that Declan tried to intimidate and rape her on two occasions, however she was able to fight him off both times. After she had kicked his arse twice, he changed his tactics.

'He told me "not everyone here can fight like you," and then warned me "if you ever stand up against me, or question my authority, you'll never know who's going to cop it,"' she explains.

Meaghan thanks Mabel for her time and she retakes her place on the bench.

Meaghan gives Declan an opportunity to respond to the sexual assault allegations and he rises from his chair. He draws himself to his full height and glares over the gathered faces. Never one to miss an opportunity for theatre or intimidation, he stands up on the log stool before addressing us.

'My friends, what this shows us is that there have been a

great many misunderstandings. While you have all been flinging allegations around, I have been trying to work out the pressing matter of who killed Tomohiro.

'While I have never really loved any of you, I would never kill one of my trusted volunteers. You elected me to protect you and that is what I have done and intend to keep doing.

'The only logical explanation, considering the sudden re-emergence of our esteemed friends, is that one of them snuck into camp and killed Tomohiro, before slinking away and hiding in the forest.'

'And why would any of us want to do that?' Meaghan asks.

'To frame me, of course,' he replies. 'You've got motivation and opportunity. If you continue to push this ridiculous case against me without considering all of the options, then you're all as dumb as I always thought you were.'

When he talks, my skin crawls and I want to jump up there and strangle him. The scariest and saddest thing is that he does have an incredible mind. He always manages to create an alternative scenario with enough credibility to put doubt into people's minds.

'And do you have any response to the other charges laid against you?' Meaghan asks.

'Not really. I admit that on one occasion, Ayesha and I got lost in the moment and I had a consensual sexual encounter with her. It was a silly mistake, but it led to the miracle of life, so I can't say that I regret it. All these other allegations are totally new to me. They appear trumped up to paint me in a bad light. I'm not guilty. No one has provided any actual evidence so let's stop wasting time on this,' Declan says.

Newton and Todd lead Declan back to his hut and after they return, we discuss the matters at hand.

'I believe that there is enough evidence to show that Declan has sustained a continued program of intimidation, including sexual assault to coerce and control others. Does anyone disagree?' Meaghan asks.

We all look to Nat, but she is staring off into the forest. I can't tell what she's thinking, but she seems to be trapped in her own thoughts.

'Raise your hand if you think Declan is guilty of the accused thirty seven counts of rape,' Meaghan asks.

Ten hands project into the air. Only Nat remains unmoved.

'That's pretty comprehensive,' Meaghan says. 'What about Tomohiro's murder?'

'My gut tells me that Declan is guilty of that too,' Ayesha says. 'But there's no evidence pinning it to him.'

'There's no evidence pinning it to anyone,' Todd adds. He looks around the group nervously, knowing that even with Declan locked up, the murderer could be within our midst.

'What about you guys?' Tiffany says, looking at me and the others. 'Declan says it could've been you.'

'Tomohiro died three days ago, right?' I ask. 'And three nights back, we had only just crossed the bridge and were still, I don't know, sixty k's away from here.'

'That's what you're saying, but why should we believe you?' Todd asks.

'Well, there's five of us that can vouch for each other,' Newton says.

'For all we know, it was two of you that did it,' Poh adds. 'Jay and Todd, for example. Playing checkers and backstabbing a lovable volunteer perhaps?'

'Stop it,' I say forcefully. 'This is what Declan wanted, for us to fight amongst one another. We have to think about this rationally.'

'We are, Sarah,' Cassidy says. 'That's the problem. There's no evidence.'

'I want to change my testimony,' Nat says suddenly.

We turn to look at her as one and she stares back gravely.

'Bring him back and I'll tell you what I know,' she says.

Declan is marched back to his stool and Nat stands facing him, her hands resting on her pregnant belly. She hesitates for a few seconds before addressing Declan directly.

'Did you mean it when you said that you've never loved any of us?' she asks.

Declan remains silent.

In that moment, Nat's composed exterior shatters. 'I love you so much, Declan. Like everyone else here, I have no idea why, but I can't stop myself from loving you.'

Nat pauses to collect her thoughts before continuing. 'On the night of Tomohiro's murder, Declan and I were in our hut as we told you. However, just before Tomohiro was killed, Declan told me he was going to the bathroom. I thought nothing of it, but when I heard Tomohiro scream, I rushed to the door. Declan appeared out of Tomohiro's hut and rushed towards me. He told me that he had found Tomohiro stabbed in the back. He seemed genuinely shaken.

'He explained that we should stay inside because everyone would suspect him if he was the first person to find Tomohiro. I believed him. I wanted to trust him so badly that I gave him my blind faith.'

Nat becomes silent. Declan glares at her.

'And what do you think now?' Meaghan asks.

'As Sarah said, he's talked about killing Tomohiro before,' Nat says. '"Euthanizing a heavy load" I think he called it. Plus, they had a huge fight the morning that Tomohiro died so…' She pauses a moment before completing her thought. 'So, I guess he probably did it.'

It's clear how much it breaks her heart to say the words out loud. For some unknown reason, she is devoted to him, but now she finds herself the first person to put a nail in his coffin.

Declan is led away. Now that he knows the odds are stacked

against him, we take no chances and tie him up securely inside his hut.

'I'm happy to deliberate further if needed,' Meaghan begins, once we've all sat back down. 'But I'll check first. In light of Nat's new evidence, who thinks Declan is guilty of killing Tomohiro?'

Seven hands shoot into the air immediately. Nat raises hers more tentatively, followed by Meaghan and Jay. Only Tiffany's hand remans unraised. We all look at her curiously and she slowly lifts a hand into the air.

'It's good to think these things through carefully,' she says with a shrug.

'Okay, so guilty on all charges. What do we do now?' Cassidy asks.

'We have to decide on a punishment,' Meaghan says.

'We can't keep him as a prisoner,' Newton says. 'There's no way we can afford to feed someone who isn't contributing to the society.'

'We could force him to work?' Jay suggests.

'That's not really punishment, though,' Poh says. 'We all have to work anyway.'

Ayesha shifts on her seat uncomfortably. 'Considering the severity of his offences and the fact that we can't send him to prison for life, there's only one appropriate punishment. The death penalty.'

Several volunteers nod but others look disgusted. The thought of ending a life on purpose, in this world where life is so precious and rare, seems preposterous. But having a convicted murderer around isn't a real option either.

'We could exile him,' I suggest. 'Banish him from camp and forbid him from using any of our resources.'

This suggestion also receives some nods of agreement but doesn't obtain overall approval.

The conversation goes around in circles with different ideas raised and rejected, but no conclusion reached.

'I have another suggestion,' I say finally, after sitting back and watching the debate unfold for a few minutes. 'Why don't we let Declan decide?'

'What!? You've got to be kidding me?' Ayesha exclaims. 'He would just choose the best punishment.'

'I'm not saying he gets to choose exactly what punishment he receives,' I explain. 'From what I can tell, we're arguing between killing him and exiling him. Let him choose which.'

'I like it,' Cassidy says.

'Makes sense to me,' Todd adds.

For what might be the final time, Jay and Todd retrieve Declan from the hut. He stands next to his log stool and, in respect of the occasion, we all stand as Meaghan announces the verdict.

'Declan McCallister. You have been found guilty of all charges laid against you. For the record, that is one charge of murder, thirty seven counts of rape and three of indecent assault. All of us gathered here today have agreed on these verdicts unanimously. We have also agreed that you can choose your punishment from the following options: complete exile from our camps and society, with only the clothes on your back and three days' worth of rations; or execution via gunshot to the head. Do you understand?'

He nods. 'But you're making a terrible mistake. I didn't kill Tomohiro.'

Meaghan ignores his final protestation. 'Can you tell us what punishment you choose?'

He takes a moment to consider before answering. 'I'm not stupid. Exile out here is also a death sentence, but a slow, painful and lonely one. If any one of you actually has the guts to pull the trigger, kill me here and now.'

He glares at the group and everyone else avoids his gaze. They glance quickly at each other, not wanting to commit the deed and wondering if their neighbour could. Declan locks eyes with me and I stare deep into his being. Behind his brazen challenge, I see a broken soul, and as he sees the steely resolve in my eyes, I see fear on his face for the very first time.

'Someone fetch me a gun,' I command loudly.

DECLAN

Declan lies in his bed unable to sleep. The fear inside him complete and unshakeable. His older brother snores beside him. Declan can never understand how Kieran is able to sleep. But at thirteen years old, maybe he's experienced the horror enough that it has become wallpaper to him. Something to expect. Something inevitable but not something to fear. At nine years old, it's not like that for Declan. The bile rises in his throat as he hears them coming. His father and his uncle. Drunk, but not drunk enough to not understand that what they do is wrong. It's exactly the same every time Uncle Arlo visits.

Declan lies in bed, unable to protect himself, the fear rising and rising as their footsteps approach.

Declan sits in the corner of the room, trying to convince himself that he fits in with the other nineteen-year-olds. They drink, they dance, and he feels alone. He tries to drink some beer but the smell of it just reminds him of his father's breath as he forced himself inside Declan. He dry retches at the memory, still clear as day in his mind's eye. He reaches for a bottle of Tequila instead, necking a few mouthfuls before someone wrenches the bottle from his hands.

'Hey, take it slow, mate!'

But he doesn't want to. He needs to drink quickly enough that he can't taste it, can't smell it. He wants to drink quickly enough to forget.

He stands up and stumbles towards the toilets, the alcohol acting fast on his virgin body. No one pays him any heed and the loneliness rises in him again. Followed closely by the fear. The fear of never being loved.

But then someone does notice him. His eyes lock with a girl on the other side of the room. The girl smiles and Declan feels something stir inside. He smiles back and adjusts his hair. She laughs lightly, a twinkle in her eye. The girl looks cute in her party get-up and with wavy orange hair.

Declan continues to the toilet, but tells himself that once he's finished, he will approach the girl. He's had just enough alcohol to lower his inhibitions.

He washes his hands and opens the door, ready to scan the room, but doesn't need to because the girl is right in front of him, smiling eagerly. She gently pushes Declan back into the bathroom and closes the door behind them.

'Hi,' Declan says with a laugh. 'What's your name?'

'Sh, sh, sh, shhh. Do you want this?' she asks.

'Yes. I think so.'

'Then stop talking.'

They fall into each other, and Declan is amazed at how good kissing feels. He's never been kissed before. They always skipped that part.

The girl presses herself against Declan, breathing heavily. She reaches for Declan's crotch, trying to find something to work, but Declan pushes her hand away.

'Can we just kiss a bit longer?' Declan asks.

Confusion flashes in the girl eyes, but then she smiles. 'Of course.'

'I'm Declan, by the way.'

'Reina.'

They continue to kiss and Declan continues to enjoy it, but he feels no movement downstairs. Reina kisses his neck and nibbles his ear. Declan tries to reciprocate but bites Reina's ear a bit too roughly.

She jerks back in pain and then smiles. 'Your first time?' she asks.

Declan nods.

'No worries. We don't need to do anything you don't want to do.'

They continue for only a few minutes longer. As they leave the bathroom, Reina squeezes Declan's hand and gives him a wink.

'How good is your memory?' she asks.

'Pretty good.'

'0491 571 266,' she says quickly. 'If it's meant to be, you'll remember.'

Two weeks later, Declan stands in the hallway, mustering up the courage to confront his father.

Declan remembered Reina's number. They've met up nearly every day since the party, but always with the same result.

He walks into the kitchen and silently takes a large knife out of the block they keep on the counter. His father sits at the table in his ridiculous green dressing gown. Declan puts the knife behind his back, hand shaking, and crosses in front of his father.

'Morning, Dec,' his father says, flashing a winning smile.

'Morning,' Declan replies coldly.

'Something the matter?'

'Don't pretend to care, Dad.'

'I care, son.'

'No. You don't. You didn't when you raped me, and you don't now.'

The colour drains from his father's face. He's used to being in control, not used to being confronted.

'But this isn't about you, Dad. I want to talk about me.'

'Okay,' he mutters. 'What is it?'

The palms of Declan's hands are sweaty but he can't wipe them dry. His grip on the knife behind his back is loosens.

'You've ruined me, Dad. Open your legs.'

'What?'

'I said, open your legs, nice and wide.'

His dad obliges. He's not wearing any underwear so Declan gets a glimpse of his wrinkled penis and scrotum before his father readjusts the dressing gown to cover himself. Bile rises in Declan's throat.

'You ruined me, Dad. I've met a girl who actually seems interested in me and nothing she does can get me hard. Nothing she does can get the memories of you and Uncle Arlo out of my head.'

Declan's hands have stopped shaking now. Adrenaline has taken hold and his whole body is wired. Trusting his aim, he slams the knife down, right between his father's legs, pinning the dressing gown to the chair. His father flinches.

Declan turns and walks away. Relishing the fear in his father's eyes, the terror. He smiles. It felt good making him feel that fear. It feels powerful.

Fear has kept Declan down his whole life, but no longer. Now he understands how he can use it.

The R.E.Generation testing is easy. Declan has spent his whole life lying to people, telling them what they want to hear. When he was a child, he told everyone he was fine. He smiled and looked like a regular child. When he was older, he pretended to like his friends, pretended that he cared for them. For the entrance exams of his Medicine and Dentistry degrees, he knew the correct answers, even if they weren't the answers he agreed with. So, when the

R.E.Generation recruiters ask him their questions he knows how he is supposed to answer.

Upon waking up, that all changed. In this new world, Declan doesn't need to pretend anymore. He can use fear and power to control the people around him. The only time he feels fear himself is when he looks at Sarah. She can't be cowed, and somehow her love seems as strong as his fear. People follow her too.

He can't understand because he thought he had it figured out. But now, as she stands over him, he begins to realise he might've made a mistake. Even now, she shows him love by choosing the kindest way to kill him. She holds a rifle to his head and after saying some legal bullshit, she pulls the trigger.

It's so quick and painless, he doesn't even hear the sound before everything goes black.

SARAH

CHAPTER 23

The moment Declan's body hits the ground, I drop the hunting rifle and collapse onto the forest floor myself.

I thought that having seen so much death, I would be able to handle it, that it wouldn't affect me. But taking Declan's life with my own hands is completely different. I want to tear them from my body and throw them away. My hands feel like evil beings with a mind of their own, capable of monstrosities.

'Sarah, Sarah…' I hear someone calling distantly.

I glance across at Declan's dead body. His arms are splayed at odd angles, which doesn't look very comfortable. I want to help him, to lay him out in a more dignified way, but I don't seem able to move.

'Up you get, very good,' I hear Cassidy say, as if through a thick fog. Something jerks at my armpit.

I allow myself to be pulled onto my feet and led to a hut. Cassidy sits me on a large box filled with pine needles and squats in front of me.

'Sarah, can you hear me?' she asks.

I nod.

'I think you may be in shock,' she continues. 'Your heart rate is rapid. I need you to calm down, for your baby's sake.'

Suddenly, my tactile senses return with magnified clarity. I can feel my heart racing in my chest, like it's trying to break out of its skeletal cage. My baby lays inside my womb, its weight pulling on my muscles and stretching my skin. I feel incredibly guilty for creating the life inside me. I'm not worthy of it. I'm a killer.

'Sarah, I need you to take some deep breaths with me,' Cassidy says.

I comply and my heart starts to calm down a little. Cassidy gives me a canteen of water to drink from and I lie down onto the pine needle bed.

'Take all the time you need, okay. Call me if you need anything,' Cassidy tells me tenderly.

I drift in and out of sleep. I dream of Declan both dead and alive. I relive the trauma of pulling the trigger and see grotesque imagery of Declan's brain and eyes leaking out of his skull and onto the forest floor. It's horrible and confronting, but somehow helps me process the whole situation.

Cassidy brings me a bowl of soup for dinner which I eat in my bed. The nightmares continue through the night, but when I wake in the morning, my clarity of mind has returned.

I head outside, nervous of how the others are going to treat me.

Newton is the first to see me and he embraces me in a huge hug.

'It had to be done,' he says quietly into my ear. 'You are so strong and incredible.'

'I don't feel it,' I reply.

'I know,' he says, drawing back and holding me by my shoulders. He stares me directly in the face. 'That's why I'm telling you. The more you hear it, the sooner you will believe it again. You're strong. You're a fighter. You're a leader. You're incredible.'

'Thank you,' I say.

He pulls me into a final embrace before letting me go.

Not everyone is as complimentary. Tiffany and Todd look at me nervously, as if expecting me to pull a gun on them at any moment. It will take a while to build their trust again.

Nat has withdrawn inside her hut and we don't see her for at least a week. Cassidy takes food to her and returns with buckets of waste that she empties in the Shit House.

Life slowly returns to some kind of normal as we plan for our journey back to Tan and the others. Unfortunately, Ayesha and Nat are too pregnant to travel, so we will have to wait a few weeks before we can get home. To complicate matters further, by the time they give birth, I will be almost eight months pregnant and unable to travel myself. The deep ache in my pelvic girdle has intensified, and every time I move I have to grit my teeth against the grinding pain.

I stress about Tan and the other volunteers waiting for our return and wondering where we have got to. As the midwife, I can't leave Ayesha and Nat, so Poh and Meaghan offer to journey back to inform them of the situation. After the next storm has safely passed, we wave them goodbye and they begin the trek back.

Camp seems lonely without them, but at least I have Cassidy and Newton to keep me company. I can't help but wonder when they will leave though, because everyone seems to leave eventually. First Ang, then Karim and Tomohiro. We left Smithy and Tan, and now Poh and Meaghan have left us. We struggle to survive, and yet one by one, the world separates us and picks us off.

Two weeks after Declan's death, Newton is the first person to bring it up with me. I'm lying with my head in Cassidy's lap and his question comes without any warning.

'Do you regret it?'

I know exactly what he's talking about, even though he hasn't provided any context.

'Yes and no,' I reply. 'Do you remember the first thing you told me the next day?' I ask.

'It had to be done,' he repeats.

'Exactly. I regret that it had to be done and I regret that Declan took us down this path. But I don't regret doing it. I wouldn't want anyone else feeling what I have to feel, so I'm glad I was the one.'

Cassidy strokes my hair gently. 'Are you alright?' she asks.

'I think so,' I say calmly.

Cassidy still has the weight of her own grief hanging around her neck, but she seems to have passed through the deepest depths, and needing to look after me seems to have helped her in a small way.

'You know what I don't understand?' she asks, tilting her head to one side. 'Why was Declan still proclaiming his innocence right up until the end? Even after he knew his fate, he told us he didn't kill Tomohiro.'

'Probably so that he could haunt us, even after he was gone,' Newton suggests. 'If we doubt that it was him, then we can't trust each other.'

'He was a scheming bastard,' I say, with more venom than I intended.

'Let's hope that we were correct,' Cassidy says.

'We were,' Newton replies confidently. 'It's over now. We can move on.'

TOMOHIRO

As soon as Sarah and the others leave, things get worse, not better.

Declan is in such a rage that Tomohiro stays in his hut, waiting out the storm. Eventually, Jay comes to see what Tomohiro is up to.

Jay is shocked to see that Tomohiro's wheelchair is gone. He helps Tomohiro outside and supports him as he limps to the central fireplace, dragging his dead leg behind him.

'Nice of you to join us,' Declan says aggressively. 'The turd stains have gone, ran away with their tails between their legs.'

'Who's gone?' Tomohiro asks innocently. 'They must've stolen my wheelchair in the night because it's gone too.'

Declan marches off to Tomohiro's hut and returns red in the face. 'They've taken a bunch of tools and supplies. You must've known,' he says jabbing a finger at Tomohiro.

'I'm a heavy sleeper, Dec. You know that.'

Declan concedes the point begrudgingly.

With only six volunteers left behind, the camp feels too big and a little overbearing. Cassidy and Sarah's small, triangular hut and the hut that the other Etas and Gammas slept in lie dormant.

Tomohiro finds it a bit amusing watching Declan storm around in anger. He's frustrated that he isn't in as much control as he thought he was and aggravated that the people he is angry at are outside of his grasp. He wants to rant and rave, but he knows he can't be too angry at the few people who remain, or he risks pushing them away too.

Declan channels his anger into a renewed push for finding more volunteers. Without a wheelchair, Tomohiro is reasonably useless, so Nat makes a new and improved wooden chair for him.

Declan bosses the volunteers around, telling Tiffany and Jay to climb all the trees near the open Eta pods and instructing Nat and Ayesha to dig out more chemo-pods from the landslide. His vigour pays off as they find three unbroken pods and he is able to wake up Mabel, Todd and Kirra.

Tomohiro tries his best to placate Declan and shield the others from him as the months slip by. He always takes the blame for any indiscretion or mistake, even if he had nothing to do with it.

As the volunteers are completing the final stretch of aqueduct, Kirra accidentally slips and smashes a board in half. Tomohiro quickly switches positions with Kirra, so that when Nat and Declan come to inspect their work, it is Tomohiro that is showered in curses.

Life goes by, but the volunteers are only surviving. Tomohiro doesn't feel like they're progressing anywhere. Only three things show that they're not just treading water.

After their initial morning sickness, Ayesha and Nat are soon unable to hide the fact that they are both pregnant. Tomohiro loves watching them learn about pregnancy together. He only wishes that Sarah was here to guide and instruct them.

After a few months, Ayesha's hidden garden produces its first edible vegetables and they all relish the fresh food. Mabel is a fantastic cook and she enjoys trying to better Nat's dishes. Nat, for her part, is up to the challenge and they both push each other to make even more delicious food.

Jay and Tiffany continue their physiotherapy with Tomohiro. Tiffany is incredibly positive and always draws Tomohiro's attention to the small improvements that he makes.

'Look, Tomohiro, you're moving your left shoulder,' she says excitedly.

Tomohiro shakes his head, sick of being told things are getting better.

Despite making progress over the last four months, his left arm and leg remain almost completely paralysed. His face muscles have shown impressive recovery and he is now able to stay standing in the one spot, but he is no closer to walking independently.

Slowly, Tomohiro comes to the realisation that this is his new normal. He will be wheelchair bound for the rest of his life and unable to participate effectively in this new society.

Tomohiro spends every opportunity he gets explaining his knowledge to the others. How to build a forge, bend hot metal, and how to avoid putting your back out while you work. Tomohiro also discusses different types of rocks, their uses and where to find the most versatile types.

Five months after the disappearance of Sarah and the others, Tomohiro feels like he has no more knowledge to share. He's given the others the best chance to survive. Luckily, he's got one more thing that he can do for them.

Without needing to talk about his expertise, Tomohiro's conversations turn to other things. Over the next month, he starts to share his feelings and fear with the others.

'I feel like Declan is breathing down my neck,' he tells Ayesha over breakfast. 'He seems to really have it in for me.'

'Don't take it personally, Tomohiro,' Ayesha replies. 'He's like that with everyone.'

'This is different, Ayesha. I don't feel safe. Maybe I'm just paranoid.'

Later in the week, Jay pushes Tomohiro's chair along the aqueduct, with Nat walking alongside. She can't help but waddle up the hill, her hands massaging and supporting the small of her back. They inspect the length of the aqueduct to make sure there is no damage or areas of weakness.

'Hey, Nat, has Declan talked to you about me?' Tomohiro asks.

'No,' Nat says, shaking her head. 'Why?'

'I don't know,' Tomohiro says, screwing up his face. 'He just seems a bit more volatile recently.'

'Really? I thought he'd calmed down a lot,' Jay says.

'Not with me. All I've tried to do is be kind to everyone, but I think Declan is worried I'm trying to undermine him,' Tomohiro says.

'And are you?' Nat asks.

'Of course not,' Tomohiro replies. 'But that's not what matters with Declan. All that matters is what he thinks.'

'Let me know if it gets worse,' Jay says. 'He's probably just playing mind games.'

Weeks later, Nat and Declan are discussing ways to find and kill more large lizards.

'There has to be more out there. The one I killed had eggs, so where is its mate?' Declan says.

'We should be trying to trap them,' Nat says. 'We build a big trap, put food inside and wham! Dinner for weeks.'

Tomohiro can't stop himself from weighing in. 'I'm not sure that we should hunt them at all,' he says, joining the conversation.

'Why's that?' Declan asks. 'Because you've gone soft?'

'No. Because we don't know enough about them,' Tomohiro replies.

'We know they taste good,' Declan says bluntly. 'And that means they're better dead than alive.'

'Why is everything so black and white with you, Declan?' Tomohiro asks.

'Because I see things more clearly than simple-minded people. I don't have doubts, because I know that I'm always right.'

'If Sarah were here, she wouldn't let you kill them,' Tomohiro says, probing a bit harder.

Declan scowls and Nat shakes her head, cautioning Tomohiro.

'You know what I think about people talking about *her*,' Declan says, his temper starting to grow. 'I'd kindly ask that you keep her out of it.'

'But she wouldn't,' Tomohiro says, barrelling on regardless. 'Sarah was gutted when you killed the first lizard and would fight tooth and nail to stop you trapping anymore.'

'What she thinks is irrelevant,' Declan snaps.

'Well, I think that what you think is irrelevant,' Tomohiro says. 'I say we leave the lizards be.'

'Well, I say we kill them, and I'm leader, so we kill them.' Declan is really riled up now and Tomohiro fears that he will lash out if he pushes the point much further. But still, he can't let it go.

'Just because you're leader, doesn't mean you get to make all the rules,' Tomohiro yells.

'That's exactly what it means!' Declan yells back. A large vein throbs in his forehead.

Tomohiro cowers away, letting Declan win yet another fight. Just before he rolls his chair away completely, Tomohiro turns around once more.

'You know, Declan, sometimes you're wrong. You're not as perfect as you think you are.' He pauses for dramatic effect. 'And also, everyone knows that Ayesha's child is yours.'

Declan lunges at Tomohiro but Nat catches his arm and holds him back. Tomohiro rolls his chair away and disappears inside his hut. He has a huge smile plastered across his face that he keeps hidden from Declan's view.

Later that night, Tomohiro sits in the doorway of his hut looking out into the camp. Waiting is the hardest part. Fear courses through his body – the fear of impending death.

Tomohiro has left his chair wedged against the wall and he sits in the dirt, watching the moon as it slowly progresses across the sky.

A beam of light stretches across the ground as the door to Declan and Nat's hut opens. The light flickers as a shadow fills the doorway and then disappears completely as the door is closed once more.

Tomohiro tracks Declan as he makes his way across camp. He disappears inside the Shit House and Tomohiro takes some deep breaths, steeling himself.

As soon as he sees the toilet door opening, Tomohiro springs into action. He drags himself across the floor of the hut and manages to pull himself to his feet, standing above his chair. He slams himself into the chair and he feels a piercing pain as the knife that he wedged into the chair hours before stabs into his liver. He stifles the scream that wants to escape his lips. He wriggles to loosen the knife from the chair's hold and crumples to the floor. The pain radiates through his body and he blinks tears from his eyes.

He kicks the wheelchair over and starts to call pitifully. 'Help, help…'

Tomohiro watches the doorway carefully, calling into the night and hoping Declan hears his pleas.

Tomohiro's vision starts to blur, but then there he is, standing in the doorway.

He sees the shock and confusion on Declan's face and he smiles weakly.

Tomohiro musters all his strength and lets out an almighty roar of pain.

The last thing he sees is fear in Declan's dark eyes. Tomohiro has done everything he can to help the other volunteers. Now they're on their own.

SARAH

CHAPTER 24

Any woman who has had a child, will probably tell you that the third trimester is the worst. And when you're surviving in a world of pine trees and mushrooms, it's almost unbearable.

I try and keep busy to keep my mind off the pregnancy symptoms, but there's not much I am able or need to do. It feels like I need to pee constantly and getting up multiple times a night is a killer for my sleep habits. Good practice for after the baby comes, though. Getting to sleep is also a struggle. I've always been a front sleeper and having to shift to sleeping on my side is a tough adjustment. All my other bodily systems seem to stop working correctly as my body focusses on forming a human in my womb. I have the worst heartburn that I've ever experienced and the constant constipation alternates between being highly annoying and actually quite painful. We don't have any medication to help with the symptoms, so I have to manage them the old-fashioned way. I eat smaller meals more regularly to limit the amount of food in my stomach and focus on drinking plenty of water. Of course, that compounds my difficulties with needing to pee all the time.

I watch as Ayesha and Nat start to swell, their hands and faces becoming puffy as their bodies focus on water retention. I guess I have that to look forward to late in my pregnancy too. I spend a lot of my time with Ayesha and Nat, talking them through plans for their labour and deliveries. Nat takes a long time to come to terms with Declan's execution. Although she didn't pull the trigger like I did, I think she feels equally responsible for his death. I thought that she might hate me, but in fact, the experience brings us closer together. Her pregnancy turns out to be just the tonic that she needs and she focuses her energies on keeping herself and her baby healthy.

'If I love this baby the right way,' she tells us, while Cassidy shows us how to weave a cot from thin branches, 'then I know it will have all of the good things about Declan, without the bad.'

'That sentiment is perfect,' I say, smiling.

Both Ayesha and Nat start to have Braxton Hicks contractions as their bodies prepare them for the extreme event to come. Ayesha gets very excited the first time it happens and I have to explain to her the difference between these contractions and the real deal.

Even so, it isn't many days before Nat informs me her contractions are getting closer and more intense. Adrenaline starts to course through my body as it gears up for action. This is what I'm trained for and it's exciting to be able to feel useful. My own pregnant belly is now very heavy and I've had ongoing backache for weeks. I've been resting to minimise my pelvic gridle pain, but nothing is going to stop me from delivering this baby.

Cassidy, nursing her own pregnant belly, helps me make Nat's bed extra comfortable. Nat has stayed in the small hut that she shared with Declan and it provides her with privacy throughout her labour.

The initial excitement and adrenaline soon fades as we wait for more action. Nat is in serious discomfort and continues to have contractions at regular intervals but her cervix dilates excruciatingly slowly. After many hours, she is still only at four centimetres. Cassidy and I help her breathe through the worst of the contractions and rub her back to try and help with her backache. We only have a limited supply of painkillers and nothing like what women used to get in hospital.

I check the baby's heartbeat regularly but there are no problems. We all just need to be patient and wait. The sky has darkened outside, and I know that we're in for a long night.

A few hours later, Newton rushes into the hut. 'I think Ayesha's in labour,' he pants.

Typical. After months of waiting, the two babies decide to come on the same night.

'Cassidy, go and check on her, please,' I ask. 'It'll be a while before she's ready. Hopefully Nat is done by then.'

Cassidy nods and ducks out of the hut. Newton steps inside and sits on the floor next to Nat. It's not ideal having Nat and Ayesha in different buildings.

'Are you able to walk, Nat?' I ask, hoping to take her to the larger hut.

She shakes her head. She already looks exhausted and she still has the hardest part to come.

I check her cervix again, but she is still only at seven centimetres.

Cassidy pops her head back into the hut.

'I think you'd better come,' she says.

I turn to follow her, but Nat grabs my hand and holds it tightly.

'You're doing great,' I tell her softly. 'And I'll be close. Everything will be fine. Newton will stay with you.'

She lets go of me and I check the foetal heartbeat once more before leaving. It seems a little elevated compared to before, but still nice and strong.

'Look after her and call us immediately if anything changes,' I tell Newton.

He looks a bit anxious at the responsibility but nods his understanding.

I waddle after Cassidy, holding my belly with one hand and supporting my back with the other.

We find Ayesha on her bed in the large Alpha hut. Mabel, Todd, Kirra, Tiffany, and Jay are gathered around excitedly.

'Mabel, please go help Newton,' I instruct. 'He's with Nat. Tiffany, stay, watch and learn. You other three, give us some space, please.' I wave my hand and the leftovers disappear to the other end of the hut.

Ayesha is panting and sweating profusely.

'Just breathe through the contractions, Ayesha,' I tell her calmly. 'Don't try and push until I tell you to.'

I check her cervix and am surprised to find it almost completely dilated and effaced.

'Looks like your baby is in a hurry to meet us,' I tell Ayesha with a smile.

A few minutes later, Ayesha is ready to start pushing and Cassidy holds her hand for support. Ayesha is an absolute champion and hardly complains about the pain, although she must be in agony. She is red, sweaty and breathing heavily, but she grits her teeth and pushes through the pain. There's nothing beautiful about labour. It's a hell of an ordeal, but usually there's something beautiful at the end.

I can see the baby crowning and soon its head is in my hands. With one final push from Ayesha, the baby's shoulders are through and I help her out the rest of the way. She has good colour and is breathing normally. Her eyes are closed, and she has a shock of dark hair on her head. Cassidy ties some fabric around the umbilical cord and cuts it neatly. I hand the precious child to Ayesha.

'She's beautiful,' she murmurs. 'Hello, baby girl. How are you?'

I smile and stand up, stretching my back. Tiffany and Jay have crept closer to get a look at the tiny baby.

'That wasn't as bad as I was expecting,' Ayesha says.

'It's not often I hear that,' I reply. 'Besides, it's not over yet. You've still got to deliver the placenta.'

At that moment, Mabel rushes in, flustered.

'You need to come quickly,' she says and disappears back out the door.

'You'll have to handle things here,' I tell Cassidy.

I follow Mabel as quickly as I can manage with my extra baby weight and painful pelvis. It makes everything twice as tiring and I'm not sure that I can handle much more excitement. All this rushing around risks bringing my own labour on early.

When I see Nat, I can immediately tell that the baby has shifted positions, now sitting lower in her pelvis.

'Have you been pushing?' I ask her.

Nat nods. 'Once or twice.'

I quickly assess the situation and a cold hand of dread sneaks up my spine.

'You're almost fully dilated, but your cervix is not thin enough yet,' I tell her. 'But now that we've started pushing we're gonna have to keep going. Your baby's heartbeat is quite faint. Are you ready for this?'

She nods. She holds up her hands, looking for support, and Newton and Mabel grab hold on either side.

Nat waits for the next contraction and screws up her face in effort as she tries to shift her baby.

Contraction after contraction, she pushes, and the baby ever so slowly moves closer to delivery. Nat begs for painkillers that we just don't have and screams in pain every time a new contraction arises. I monitor the baby's heart rate and after thirty minutes, it starts to slow drastically.

'Newton, can you get Cassidy please?' I ask.

Fear flashes in Nat's eyes, but I instruct her to focus on the contractions.

When Cassidy arrives, we talk in hushed whispers in the corner of the hut.

'Can you do a caesarean?' I ask her.

She shakes her head. 'The risk to Nat is too great. Better to lose the baby than lose her.'

'Shit. We need this baby out now,' I say agitated.

Cassidy checks on the baby's heart rate while I check to see if the baby is crowning yet. Cassidy gives an almost imperceptible shake of her head.

'Alright, Nat,' I say rousingly. 'I know you're exhausted. I know you've been giving it your all, but now we need you to give us even more. One hundred and twenty percent. A few big minutes and we'll have your baby out.'

She screws up her face in concentration and pain. I see the baby's head appear during the contraction, and it disappears when she relaxes.

'That's good Nat, keep it up.'

A few contractions later and the baby's head remains visible after the contraction has finished.

'She's crowning. You're almost there Nat, keep going.'

The baby's entire head emerges with the next contraction and I have to unwrap the cord that is loosely wrapped around her neck. Her face is discoloured and she is motionless.

'Final stretch, Nat! Give me a big one,' I call out.

At long last, Nat's baby is finally delivered and I hold her gently in my arms. Cassidy cuts the umbilical cord and I lay her on the foot of the bed, rubbing her with some rough cloth. She isn't breathing and she has no heartbeat.

'My baby, I want to hold my baby,' Nat cries.

'We need to check her over first,' I tell her.

Cassidy gently blows air into the baby's lungs and starts compressions using two fingers.

'What are you doing?' Nat asks in a panic.

'Your baby isn't breathing. We're trying to resuscitate her.'

'Oh god, no. No, no, no, no, no,' Nat wails.

Her body is wracked by another contraction.

'What the fuck is this?' she screams.

'You need to deliver the placenta,' I tell her calmly. 'Next contraction, push as hard as you can.'

Cassidy continues to give CPR and I try every trick that I know to revive the baby, but she doesn't respond.

Cassidy wraps the lifeless baby in a cloth and hands it to Nat.

'I'm sorry, Nat, we did everything we could,' I say. I want to add more, but no words can explain away the pain. No justification can bring Nat's precious child back to life.

Tears stream down Nat's face as she holds the baby close to her. 'Esme,' she says. 'That's your name. Esme. My beautiful baby girl.'

She holds her baby tight as she has more contractions. About thirty minutes later, she delivers the placenta and can finally stop her ordeal. She lies in silence, staring down at her baby. I can tell that she has blocked the rest of us out. It's just the two of them in her world of sorrow.

'You have a small tear,' Cassidy tells Nat. 'I'm going to sew it up for you.'

She doesn't reply, so Cassidy shrugs and sets to work anyway.

I leave them to it and waddle across the camp to check on Ayesha and her baby.

The mood in the other hut couldn't be more different. The volunteers are gathered around Ayesha and talking excitedly. Kirra has Ayesha's baby in her arms, and she is singing to her softly.

They turn to look at me as I enter and all I need to do is shake my head.

A stunned silence settles over the group. We all knew having children in this new age, without our old technology, was going to be a risky business, but the loss of Nat's baby hits everyone hard. Still, the fact that both mothers survived is its own little miracle. In this old-fashioned, new world, childbirth is risky for child and mother alike.

'What happened?' Tiffany asks quietly.

'Hard to say,' I reply. 'It was a very long labour. I assume the cord was compressed and we couldn't get her out quickly enough.'

'Oh, Sarah, I'm so sorry,' Ayesha says.

'It's not me who needs your sympathy. Poor Nat, I don't know how she's going to get through this.'

At that moment, Ayesha's baby let's out an adorable hiccup, followed by a gurgle. I look at her lovingly.

'Have you decided on a name?' I ask Ayesha.

'For the first child born in the new world? It's a lot of pressure,' she replies. 'But I decided on Violet. One of the first flowers to bloom after winter.'

I nod approvingly.

The news of Nat's loss has dampened everyone's spirits, and the lateness of the night starts to catch up with everyone. We're well into the wee hours of the morning.

The others retire to their beds, Violet nestled in her mother's arms.

I check on Nat before taking myself to bed. She also sleeps with her baby resting against her chest, a protective arm holding the lifeless babe close. In sleep, Nat looks calm but come morning, she's going to have to face reality all over again.

I leave her alone and back in my own hut that I share with Cassidy and Newton, I try to find a comfortable position to get some rest.

After our late night, I stay in bed until late the following morning, but eventually I'm awoken by Todd knocking loudly on the wall above my head.

'Alright, alright,' I mumble.

'I can't find Nat,' Todd says anxiously.

I shake my head, trying to force myself to wake up more quickly and sit up in bed. Cassidy and Newton roll over in their beds to listen in.

'What do you mean?' I ask hazily.

'I can't find her anywhere in camp,' he replies.

I throw on my clothes and follow Todd into the midmorning sun. Jay is trying to light a fire at the central fireplace, but there are no other signs of life.

I check Declan and Nat's hut. Esme has been lain gently in the middle of the pine needle bed. I quickly search the room to see if anything else is amiss but find no clues. Nat has taken her clothes but has left the rest of her personal effects.

Cassidy and Newton have roused the other volunteers and we gather at the fireplace. Ayesha carries Violet in her arms, looking every bit a mother.

'Nat seems to have taken herself off somewhere,' I tell the others. 'Given her ordeal last night, I can't imagine she is in a very good way. We need to find her and bring her back to camp. Does anyone know where she might be?'

'There is a particular section of forest that is her favourite for cutting down trees,' Tiffany says.

'She used to help me at the secret garden sometimes,' Ayesha says.

'Okay, Tiffany, Cass, you check those places out,' I say. 'Newton, follow the aqueduct right up to the spring. She may have gone there for some reason. Ayesha, you stay here. Everyone else, fan out from camp and see if you can find anything.'

I retrieve Aravis from her hut. She and Tumnus have taken over the hut that Cassidy and I used to occupy, where they were originally hatched.

'Can you find Nat?' I ask her, not expecting anything.

She trots over to Nat's hut and flicks her tongue out, smelling the air. She continues past the hut and into the forest. I've learnt not to ignore Aravis, so I follow her, while doubting we're going to find anything.

I waddle through the trees, scanning the ground for any trace of Nat. Aravis stops and I look at her, confused. She looks up into the branches above our heads and I follow her gaze.

Swaying amongst the pinecones, Nat's body hangs from a large branch. The rope tied around her neck digs into her skin and her eyes stare blankly into the treetops.

I dry retch and double over. When will this world stop taking from us? I shake my head, knowing the answer already. Never.

Aravis rubs her head against my hip, recognising my pain and sharing her empathy. I stroke her neck and look back up at Nat. She was never my favourite fellow volunteer, but over the month, she progressed to being a friend.

I turn my back on her and walk back to camp. Once some of the others return, I lead them to Nat's body. There's no graceful way to get her down, so Todd climbs the tree and cuts the rope. Nat's body tumbles from the tree, smacking branches on its way to the forest floor.

Kirra lets out a little squeal and Cassidy turns away in horror.

Todd climbs back down to us and we lift Nat onto our shoulders. We carry her gently back to camp and lay her on the ground.

'The first suicide of the new society,' Cassidy whispers.

'She went through hell,' Kirra says.

'First Declan and then her baby,' I say. 'I should've seen this coming. I should've looked after her better.'

'You couldn't have known,' Cassidy says. 'None of us could.'

We fall into silence and sit next to her body, waiting for the others to return. Everyone is shocked and dismayed at the sight they find when they enter camp. Tiffany breaks down sobbing and

Mabel holds her in a tight embrace, tears welling in her eyes as well.

I collect Esme from Nat's hut and once Newton returns from the spring, we carry them both into the tree line. We find the wooden cross over Declan's grave and all contribute to digging a shallow hole next to it.

We place Nat in the grave with her baby resting in her arms. Cassidy closes Nat's eyes and she looks at peace. Almost like she looked when I saw her sleeping the previous night.

We say a few words of farewell and shovel the dirt on top of Nat, leaving her to her final resting place.

Everyone is edgy over the next month and a half as we wait for my baby to arrive. We're keen to get back to the others and happy to leave this camp behind, but my pelvic girdle pain is so extreme that we are stuck for the moment. Even being wheeled around in Tomohiro's chair sends pain lancing through my back and upper legs, so I am confined to bed rest.

Violet is the shining light in our otherwise gloomy and anxious days. Jay calls her Spud and I can't argue the sentiment, because she doesn't do much. Every time that I see her, she reminds me that there is also good in this world and a hope for our future. Ayesha is a natural mother and cares for Violet with ease and love.

I continue to have difficulty sleeping, struggling to get comfortable with my massive baby bump. The only consolation being that I dream less and therefore don't have to confront Declan in my nightmares. I'm fascinated to watch the changes in my body firsthand, as my due date draws near. I develop varicose veins and can't help but loath my sausage fingers. My practice contractions begin and a nervous excitement fills my body. I have to wait a further week before the contractions start coming with regularity and I know I am heading into labour.

I call Cassidy and Tiffany and they fuss about me as I breathe through my early contractions. Having delivered hundreds of babies, I thought that I was going to be prepared for my own labour, but I couldn't have been more wrong. I feel like I have lost control of my body as brand new physiological processes take over. I move constantly, trying to find a comfortable position, but nothing about this is comfortable. I take some tablets, but they do nothing for the pain. The agony is unbelievable, and I have a new level of respect for the women I cared for previously who refused painkillers. I wish I had that choice.

Cassidy and Tiffany keep me calm and Newton appears beside me to hold my hand.

My labour is almost as long as Nat's and I begin to worry. After a final few minutes of pushing, during which I defecate myself, I finally feel some relief as Cassidy helps my baby fully into the world.

I take a giant breath, feeling exhausted. Tiffany cuts the cord and my baby starts to cry. The sound of their little voice instantly brings tears to my own eyes.

'Are they alright?' I ask.

'Everything seems perfect,' Cassidy says, wrapping my baby in a cloth.

'Is it a boy or girl?' I ask.

'A beautiful baby girl,' Cassidy replies handing her over to me.

She stops crying and I smile down at the fragile little thing. I stroke her cheek and she turns towards my breast. After a short period of rooting, she finds my nipple, latches and takes her first colostrum.

'Have you had time to think of a name?' Tiffany asks.

I nod. 'Tan and I talked about it before we left. Kari. In honour of Karim. May you grow to be as strong and caring as your namesake, my little darling.'

Cassidy wipes a tear from her eye and I hold her hand softly.

I wince as my uterus contracts painfully. 'Here we go again,' I say.

'Twins?' Tiffany asks.

'Let's hope not,' I say. 'My placenta, hopefully.'

A few minutes later, I birth my placenta and collapse my head onto the bed, feeling weak.

'Is it supposed to look like that?' Tiffany asks.

I try and lift my head to see what she's talking about but can't muster the energy.

'No, it should be bigger,' Cassidy says. 'Sarah, I think some of your placenta is retained.'

I groan.

'Fuck, that's a lot of blood,' Tiffany says.

'Tiff, get me as many blankets as you can. I'm gonna have to try and get the rest of the placenta out.'

I start to feel woozy and I can feel my breathing getting faster.

'Stay with me, Sarah. I need you to promise you're gonna stay with me.' Cassidy's voice is level, but I hear a hint of panic behind the composure.

'I'm not sure I can make that promise,' I say weakly. 'Make sure you take Kari to Tan. She has his ugly nose.'

'Don't be ridiculous. You can take her to him yourself.'

She looks at me and I smile feebly. She shakes her head, refusing to believe the undeniable truth.

I want to apologise for abandoning her, but I don't have the energy.

She tries to stem the flow of blood, but it isn't long before the world turns foggy, and then grey. Soon, there is no colour at all.

Acknowledgements

When in the throes of writing a manuscript it can often feel very lonely, but in reality it takes a village to raise a novel.

I can't thank my wife, Georgie, enough for all the love, support and time she has given me to get this project across the line. She patiently prodded me to keep writing, accepting that sometimes I would disappear for hours into my world of pine trees and mushrooms. The last year she has been superhuman raising our daughter and giving me time and space to edit and edit and edit.

To my family who always made me feel special. You believed in me from day dot, and that has lead to many wonderful things. And Ben, you know what you did.

Thanks to Shandar, Ryan, Peter, Chris, Rosie, Jane, Polly and Barney, who read my book before it was polished and still told me it was good and worth fighting for. Your ideas encouragement and enthusiasm spurred me to keep pushing, and look at where we are now.

To Anna and the team at MidnightSun Publishing, there are not enough words in the world to express my gratitude. You have literally made my dream come true, and there is no greater gift a human can give.

To you the reader, thank you for reading my passion project, especially if you are even reading these acknowledgments. I hope it made you think and maybe even cry.

I would also like to acknowledge victims of sexual and domestic violence. I am sure that if you managed to read this book that it poked the wound, and I am sorry for that. But I want you to know, it isn't your fault and you are mighty. You are stronger and far more impressive than any of the cowards who perform such horrendous acts.

Finally I want to pay my respects to all the mothers out there and to all of the children who didn't make it earthside. To our two lost pregnancies, you are always in our hearts and minds. And baby Xander, you are beautiful and your parents are blessed to have held and loved you.

Treat yourself, your friends, your family and your world with love.

About the Author

Michael has been writing plays and stories, and developing board games for as long as he can remember. *R.E.Generation* is his debut novel, but he already has a number of others in the works. Michael holds a Bachelor of Science (majoring in Anatomy) and a Masters of Speech Pathology. He has worked as a paediatric speech pathologist for over 5 years. Michael and his wife are passionate about taking small steps to minimise the impact humans leave on our wonderful planet. Michael also has a keen interest in theatre, sports, mathematics, puzzles and snorkelling. He lives on the outskirts of Melbourne with his wife, Georgie, and their young daughter.